D0687614

BOUNDARIES AND BARRIERS

BOUNDARIES AND BARRIERS

On the LIMITS to Scientific KNOWLEDGE

Edited by

John L. Casti
Santa Fe Institute
Santa Fe, New Mexico, U.S.A.

and

Anders Karlqvist
Royal Swedish Academy of Sciences
Stockholm, Sweden

Addison-Wesley Publishing Company, Inc.
The Advanced Book Program

Reading, Massachusetts · Menlo Park, California · New York
Don Mills, Ontario · Harlow, England · Amsterdam · Bonn
Sydney · Singapore · Tokyo · Madrid · San Juan
Paris · Seoul · Milan · Mexico City · Taipei

Library of Congress Cataloging-in-Publication Data
Boundaries and barriers : on the limits to scientific knowledge /
edited by John L. Casti and Anders Karlqvist.
p. cm.
"Advanced book program."
Includes bibliographical references and index.
ISBN 0-201-55570-0
1. Science—Philosophy. I. Casti, J. L. II. Karlqvist, Anders.
Q175.B738 1996
501—dc20 96–28594
CIP

Cover design by Suzanne Heiser
Typeset by the editors in Times Roman

1 2 3 4 5 6 7 8 9—MA—0099989796
First printing, August 1996

Contents

Introduction

Beginning in 1983, the Swedish Council for Planning and Coordination of Research has organized an annual workshop devoted to some aspect of the behavior and modeling of complex systems. These workshops are held in the scientific research station of the Royal Swedish Academy of Sciences in Abisko, a rather remote location far above the Arctic Circle in northern Sweden. In May 1995, during the period of the midnight sun, this exotic venue served as the gathering place for a small group of researchers from across a wide disciplinary spectrum to examine the question of the limits to scientific knowledge.

As one wanders through the triumphs of 20th-century science, certainly one of the most striking aspects is the uncommon number of major results that can only be described as *limitative.* Einstein's special and general theories of relativity both rely on the imposition of a cosmic speed limit to the velocity of any material object, Heisenberg's Uncertainty Principle imposes limits on how accurately one can observe conjugate attributes of an object, like its position and momentum, and even in the world of pure mathematics the work of Gödel and Turing sparked off a plethora of results telling us that there are even limits to the truth-generating power of deductive logic itself.

This work by Gödel and Turing can be regarded as saying that there are perfectly sensible, even simple, questions about numbers that no system of human logic will ever enable us to answer; they are beyond the bounds of human reasoning. As IBM researcher Gregory Chaitin once expressed it, the world of arithmetic is just too complex for us. The 1995 Abisko meeting was convened to ask the same question about the world of nature and humans: Is the *real world* too complex for us? Are there questions in the areas of physics, biology, and/or economics that are beyond the power of science to ever answer?

Note carefully that what's involved here is whether there are *logical* barriers to answering such questions. It's manifestly clear that there certainly are "unanswerable" questions in every field of life. But generally these questions—such as the existence of life forms in Andromeda or the shortest route visiting all the world's capital cities—are unanswerable because we don't have the time, money, energy, and/or political will and desire to answer them. They are in no way logically impossible to answer—just very difficult or practically impossible. But the participants at the Abisko meeting were after much bigger game; the logical, not the practical, limits of scientific knowledge.

At the end of the meeting, all participants agreed to prepare chapter-length discussions of their views on this matter of logical limits to science. This volume is the outcome of that decision. As can be seen from a quick glance at the Table of Contents, the contributions tend to center on the issue of limits in physics and biology, tied together with some more philosophical speculations rooted in computer science and mathematics. The provisional message coming out of these disparate efforts is that, unlike mathematics, there is no knock-down, airtight argument to believe that there are questions about the real world that we cannot answer—in principle. The chapters that follow provide the arguments supporting this encouraging, but fairly outrageous, claim.

John Casti, Santa Fe
Anders Karlqvist, Stockholm

Contributors

John D. Barrow—Astronomy Centre, University of Sussex, Brighton BN1 9QH, UK

Leo W. Buss—Departments of Biology and Geology & Geophysics, Yale University, New Haven, CT 06520-8104 (email: `buss@muggiea.biology.yale.edu`)

John L. Casti—Santa Fe Institute, 1399 Hyde Park Road, Santa Fe, NM 87501 (email: `casti@santafe.edu`)

N. C. A. da Costa—Research Group on Logic and Foundations, University of São Paulo, 05655–010 São Paulo, Brazil

F. A. Doria—Research Center on Mathematical Theories of Communication, Federal University of Rio de Janeiro, 25660–020 Petropolis RJ, Rio de Janeiro, Brazil (email: `doria@omega.lncc.br`)

Walter Fontana—Institute of Theoretical Chemistry, University of Vienna, Währingerstraße 17, 1090 Vienna, Austria (email: `walter@tbi.univie.ac.at`)

James B. Hartle—Department of Physics and Institute of Theoretical Physics, University of California, Santa Barbara, CA 93106-9530 (email: `hartle@cosmic.physics.ucsb.edu`)

Piet Hut—School of Natural Sciences, Institute for Advanced Study, Princeton, NJ 08540 (email: `hut@guiness.ias.edu`)

Harold J. Morowitz—Krasnow Institute for Advanced Study, George Mason University, Fairfax, VA 22030 (email: `morowitz@gmu.edu`)

Robert Rosen—Department of Physiology and Biophysics, Dalhousie University, Halifax, Nova Scotia B3H 4H7, Canada

Karl Svozil—Institut für Theoretische Physik, Technical University of Vienna, Wiedner Hauptstraße 8-10/136, 1040 Vienna, Austria (email: `svozil@tph.tuwien.ac.at`; www: `http://tph.tuwien.ac.at/~svozil`

Joseph F. Traub—Computer Science Department, Columbia University, New York, NY 10027 (email: `traub@cs.columbia.edu`)

Chapter 1

LIMITS OF SCIENCE

JOHN D. BARROW

I. Introduction

An apocryphal story is often told of the patent office whose director made an application for his office to be closed at the end of the nineteenth century because all important discoveries had already been made [1]. Such a fable nicely illustrates the air of overconfidence in human capabilities that seems to attend the end of every century. In science this confidence often manifests itself in expectations that our study of some branch of nature will soon be completed. Typically, there is great confidence in the scope of a successful line of enquiry, so much so that it is expected to solve all problems within its encompass. But the quest to complete its agenda often uncovers a fundamental barrier to its completion – an impossibility theorem.

I want to draw attention to some of the barriers to scientific progress that may be encountered in the future – some of which have been encountered already. They are interesting, not because of some satisfaction at seeing science circumscribed, but because they involve key ideas in science that are as important as new discoveries. As a particular example, I shall focus the discussion primarily upon the search for a "Theory of Everything" – by which particle physicists mean simply a unified description of the laws governing the fundamental forces of Nature [2]. I want to consider four general types of limitation which could prevent the completion of our investigations into the form of a Theory of Everything.

II. Existential Limits

The first question we should pose is whether such a Theory of Everything exists at all. It is possible that some of our research programs are directed at discovering theoretical structures that do not exist. At root this is just the old philosophical problem of distinguishing between knowledge of the world and knowledge of our mental models of the world. Limitations upon our abilities to understand fully the latter might be best interpreted as "limits of scientists" rather than "limits of science." Bearing this distinction in mind we should be sensitive to the sources of some of our favorite scientific concepts.

The notion that it is more desirable to seek a description of the Universe in terms of a single force law, rather than in terms of 2, 3 or 4 forces, is at root, like the entire concept of a law of nature, a perspective that has clearly definable religious origins [2]. Although for a number of years there has been growing interest in the structure of superstring and conformal field theories as candidates for a Theory of Everything [3] these have proven too difficult to solve. Hence, there are as yet no predictions, observations, or tests to decide whether or not this line of inquiry is in accord with the structure of the world, rather than merely a new branch of pure mathematics.

There have been some studies of particle physics theories in which there is no unification of the fundamental forces of Nature at very high energies. So-called "chaotic gauge theories" [4] assumed that there were no symmetries at all at very high energy; these emerged only in the low-temperature limit of the theory, a limit that necessarily describes the world in which atom-based life forms live. These ideas have not yet been thoroughly investigated, but they are similar in spirit to some of the recent quantum cosmological studies on the nature of time. They argue that time is a concept that only emerges in the low-temperature limit of a quantum cosmological theory, when the universe has expanded to a large size compared to the Planck length scale [5].

Even if one concedes that there is a Theory of Everything to be found, it should not be assumed that it is logically unique. When the study of superstring theories began in the early 1980s, it appeared that there might be just two of these theories to choose from [6]. At first in these theories, logical self-consistency appeared to be a much more powerful restriction than had previously been imagined. As investigations have proceeded, a new perspective has emerged. There may be many thousands of possible string theories. While it may ultimately transpire that these theories are not as different as they first appear, many of them may turn out to be just different ways of representing the same underlying theory. If they are distinct then the message may be that there exist many different self-consistent Theories of Everything.

Some of them may contain the four forces of Nature that we observe, while others may lack some of these forces—or contain additional forces.

While we know that we inhabit a world described by a system of forces that permits the existence and persistence of stable complex systems—of which DNA-based "life" is a particular example—if other systems of forces also permitted complex life to exist, then we would need to explain why one logically consistent Theory of Everything is to be chosen rather than another.

III. Conceptual Limits

If a deep Theory of Everything does exist, then how confident should we be about our ability to comprehend it? This depends upon how deep a structure it is. We could imagine an infinitely deep sequence of structures that we could only ever partially fathom. Alternatively, the Theory of Everything may lie only slightly below the surface of appearances and be well within our grasp to comprehend. It does not follow that the most fundamental physical laws need be the deepest and most logically complicated aspects of the universal structure.

In practice, we have learned that the outcomes of the laws of nature are invariably far more complicated than the laws themselves because they do not have to possess the same symmetries as the laws themselves [2]. However, we must appreciate that the human brain has evolved its repertoire of conceptual and analytical abilities in response to the specific challenges posed by the tropical savanna environments in which our ancient ancestors developed [7] over half a million years ago. There would seem to be no evolutionary need for an ability to understand elementary particle physics, black holes, and the ultimate laws of nature. Indeed, it is not even clear that something as simple as rationality was selected for in the evolutionary process. This pessimistic expectation could be avoided if it is true that the laws of nature can be understood in full detail by a combination of very elementary concepts—like those of counting, cause-and-effect, symmetry, and so on—conceptions that do seem to have adaptive survival value. In that case, our scientific ability should be seen as a by-product of adaptations to environmental challenges that may no longer exist, or which are dealt with in other ways following the emergence of consciousness. Moreover, much of our most elementary intuition for number and quantity may be a by-product of our linguistic instincts. It might even be that the structure of counting systems in many primitive and ancient cultures, together with notions like the place/value notation [8], derive from our complex genetic programming for language acquisition. Our linguistic abilities are far more impressive than our mathematical abilities, both in their complexity and their universality among humans of all races.

We might ask whether a Theory of Everything will be mathematical. All our scientific studies of the universe assume that it is well described by mathematical structures. Indeed, some would say much more: that the universe *is* a mathematical structure [8]. Is this really a presumption? We can think of mathematics as being the description (or the collection) of all possible patterns. Some of these patterns have physical manifestations, while others are more abstract. Defined in this way, we can see that the existence of mathematics is inevitable in universes that possess structure and pattern of any sort. In particular, if life exists then pattern must exist, and so must mathematics. There is at present no reason to believe that there exists any type of structure that could not be described by mathematics. But this does not mean that the application of mathematics to all structures will prove fruitful. Indeed, the point is tautological: given another type of description, this would simply be added to the body of what we call mathematics.

Let us return to the issue of our evolutionary development, and the debt we owe to it. One way to look at the evolutionary process is as a means by which complex ("living") things produce internal models, or bodily representations, of parts of the environment. Some aspects of the astronomical universe—like its vast age and size—are necessary prerequisites for the existence of life in the universe. Billions of years are required for nuclear reactions within stars to produce elements heavier than helium, which form the building blocks of complexity. Because the universe is expanding, it must also be at least billions of light-years in size. We can see how these necessary conditions for the existence of life in the universe also play a role in fashioning the view of the universe that any conscious life-forms will develop.

In our own case, the fact that the universe is so big and old has influenced our religious and metaphysical thinking in countless ways. The fact that the universe is expanding also ensures that the night sky is dark and patterned with groups of stars, and this feature of our environment has played a significant role in the development of many mythologies and attitudes towards the unknown [7]. It has also played an important role in the course of evolution by natural selection on the earth's surface. Many of the general ideas that we have about the nature of the universe, and its origins, have their basis in these metaphysical attitudes, which have been subtly shaped within our minds by the cosmic environment about us.

IV. Technological Limits

Even if we were able to conceive of and formulate a Theory of Everything—perhaps as a result of principles of logical self-consistency and completeness alone—we would be faced with an even more formidable task: that of testing it by experiment. There is no reason why the universe should have been constructed for our convenience. The decisive features predicted by such theories might well lie beyond the reach of our technology. We appreciate that this is no idle speculation. In recent years the US Congress has cancelled the SSC project on the grounds of its cost, and the future of experimental particle physics now rests in the hands of the LHC project planned for CERN. These projects seek to increase collider energies far into the TeV range to search for evidence of supersymmetry, the top quark and the Higgs boson—all pieces in the standard model of elementary particle physics [9]. However, even the energies expected in the LHC fall short by a factor of about one million billion of those required to test directly the pattern of fourfold unification proposed by a Theory of Everything. Restrictions of economics and engineering, the pressing nature of other more fruitful and vital lines of scientific inquiry, and the collective wishes of the voters in the large democracies, doom such direct probes of the ultra-high energy world.

Unfortunately for us, the most interesting and fundamental aspects of the laws of nature are intricately disguised and hidden by symmetry breaking processes which occur at energies far in excess of those experienced in our temperate terrestrial environment. The laws of nature are only expected to display their true simplicity at unattainably high energies. According to some superstring theories, the situation is even worse: the laws of nature are predicted only to exhibit that simplicity and symmetry in more space dimensions than the three that we inhabit. The universe may have nine or twenty-five spatial dimensions. But only three of those dimensions are now large and visible; the rest are confined to tiny length-scales, far too small for us to scrutinize directly [3, 10]. Our best hope of an observational probe of them is if there are processes which leave some small, but measurable, traces at very low energies.

Alternatively, particle physicists look increasingly at astronomical environments to produce the extreme conditions needed to manifest the subtle consequences of their theories. In principle, the expanding universe experienced arbitrarily high energies during the first moments of its expansion. The nature of a Theory of Everything would have influenced the character of those first moments and may have imprinted features upon the universe that are observable today [11]. However, the most promising theory of the behavior of the very early universe—the inflationary universe—dashes such hopes. Cosmological inflation requires the universe to have undergone a brief period of accelerated expansion during its early expansion history [12]. This has the advantage of endowing the universe with various structural features that we can test today. However, it has the negative effect of erasing all information about the state of the Universe before inflation occurred. Inflation wipes the slate clean of all the information the universe carries about the Planck epoch when the Theory of Everything would have its principal effects upon the structure of the universe [11].

Limits upon the attainment of high energies are not the only technological restrictions upon the range of experimental science. In astronomy, we appreciate that the bulk of the universe exists in some dark form whose existence is known to us only through its gravitational effects [13]. Some of this material is undoubtedly in the form of very faint objects and dead stars, but the majority is suspected to reside in populations of very weakly interacting particles. Since there are limits on our abilities to detect very faint objects and very weakly interacting particles in the midst of other brighter objects, and strongly interacting particles, there may be a technological limit upon the extent to which we can determine the identity of the material content of the universe. Likewise, the search for gravitational waves might, if we are unluckily situated in the universe, require huge laser interferometers or space-based probes in order to be successful, and might therefore be severely limited by costs and engineering capabilities in just the same way that high-energy physics has proved to be.

At present we know of four forces of nature, and attempts to create unified theories of physics which join them into a single superforce described by a Theory of Everything assume that only these four forces exist. But there could exist other forces of nature, so weak that their effects are totally insignificant both locally and astronomically, but whose presence in the Theory of Everything is crucial for determining its identity. There is no reason why the forces of nature should all have strengths that are such that we can detect them with our present technology.

It is interesting to recall how accidents of our own location in the universe have made the growth of technical science possible in many areas. Even if some hypothetical extraterrestrial life-form required high intelligence to survive in its local environment, it should not be assumed that this means that they will have highly developed scientific knowledge in all areas.

For example, it is an accident of geology that our planet is well endowed with accessible surface metals. Without them, no technology would have developed. The existence of the Earth's magnetic field, together with the presence in the Earth's crust of magnetic and radioactive materials, has led to our understanding of these forces of nature. Similarly, an accident of meteorology has saved the earth from a permanent sky-covering of cloud, and enabled us to develop astronomical understanding. And, even with cloud-free skies, we have benefited from an accident of astronomy: our particular location in the disk of the Milky Way could easily have been shrouded by dust in all directions (rather than just in the plane of the Milky Way), so inhibiting the development of optical astronomy. Our ability to test Einstein's theory of gravitation hinged originally upon two coincidences about our solar system: The fact that the full Moon and the Sun have the same apparent size in the sky (despite being very different in real size, and in distance from us) means that complete eclipses of the Sun can be seen from earth. This enabled the light-bending predictions of Einstein's general theory of relativity to be tested in the first half of this century. Similarly, the presence of a planet with an orbit as close to the Sun as Mercury's enabled the predictions of perihelion precession of planetary orbits to be checked. Without these fortunate circumstances scientists would have been left with a largely untestable theory. These examples are given simply to make the point that scientific progress is not necessarily an inevitable march of progress that will occur in any civilization that is "advanced" by some criterion. There may be accidents of environment that prevent the development of science in some directions, while facilitating it in others.

V. Fundamental Limits

One hallmark of the progress of a mature science is that it eventually begins to appreciate its own boundaries. In the present century we have seen many examples of this. In an attempt to extend a theory in new ways what has often been discovered is some form of "impossibility" theorem—a proof using the assumptions of the theory that certain things cannot be done or certain questions cannot be answered. The most famous examples are Heisenberg's Uncertainty Principle in physics,

Einstein's speed-of-light limit on signaling velocities, Gödel's incompleteness theorem in mathematical logic, Arrow's theorem in economics, Turing uncomputability, the intractability of NP-complete problems, like the "Traveling Salesman Problem," and Chaitin's Theorem about the unprovability of algorithmic randomness. It may transpire that these impossibility results, together with many others that are suggested by studies of space-time singularities, space-time horizons, and information theory, may place real restrictions on our ability to frame or test a Theory of Everything.

We know already that the finite velocity of light ensures that we have a visual horizon (about fifteen billion light years away) in the universe, beyond which light has not had time to reach us since the expansion of the Universe began. Thus, we are always prevented from ascertaining the structure of the entire universe (which may be infinite in extent). Astronomers are confined to studying a finite portion of it, called the visible universe. It may be that the visible universe does not contain enough information to characterize the laws of physics completely. Certainly, it does not carry enough information to determine the nature of any initial state for the whole universe without the introduction of unverifiable, and necessarily highly speculative, "Principles" to which the initial state is believed to adhere [11, 14]. Many such Principles have been proposed—the "no boundary condition" of Hartle and Hawking [15], the minimum gravitational entropy condition of Penrose [16], and the out-going wave condition of Vilenkin [17] are well known examples. Global principles of this sort will all provide quantum-averaged specifications of the entire cosmological initial state. But our visible universe today is the expanded image (possibly reprocessed by inflation) of a tiny part of that initial state, where conditions may deviate from the average in some way, if only because we know that they happened to satisfy the stringent conditions necessary for the eventual evolution of living complexity.

As yet Gödelian incompleteness has not made any restriction upon our physical understanding of the universe, although it has the scope to do so because it has been shown by Chaitin [8, 18] that it can be recast as a statement that sequences cannot be proved to be random (a Gödel undecidable statement might be just the one needed to characterize the order in a sequence). Gödel's theorem requires that logical systems large enough to contain the whole of arithmetic are either inconsistent or incomplete. Now although modern physicists use the whole of arithmetic (and much more besides) to describe the physical universe, we cannot conclude from this that there will exist some undecidable proposition about the universe. The description of the laws of nature may require only the decidable part of mathematics. Alternatively, it may not be necessary to use a mathematical structure as rich as the whole of arithmetic to characterize the universe. The fact that we do may simply be because it is quick and convenient to do so, or because of a sequence of historical accidents that have bequeathed a particular mathematical system. If one uses a smaller logical system than arithmetic (like euclidean geometry, or Presburger arithmetic, with only + and – operations) then there is no Gödel incompleteness: in these simpler axiomatic systems all statements can be demonstrated to be true or false, although the procedure for deciding may be very lengthy and laborious.

Recently, examples have arisen in the study of quantum gravity [19], in which some quantity that is observable, in principle, is predicted to have a value given by a sum of terms that is uncomputable in Turing's sense (the list of terms in the sum could only be provided by a solution of a classic problem which is known to be uncomputable—the list of all compact four-manifolds). Again, while this may have fundamental significance for our ability to predict, we cannot be sure that it imposes an unavoidable restriction. There may exist another way of calculating the observable in question using only conventional computable operations. Another interesting example of this sort arises in the biochemical realm, where it is known that nature has found a way to fold complex proteins quickly. If it carries out this searching process by the process that we would use to program it computationally, then it would seem to be uncomputable in the same way that the "Traveling Salesman Problem" is uncomputable. However, nature has found a way to carry out the computation in a fraction of a second. As yet we don't know how it's done. Again, we are led to appreciate the gap between the limitations on nature and the limitations upon the particular mathematical, computational or statistical descriptors that we have chosen to describe its actions.

VI. Summary

In this brief survey we have explored some of the ways in which the quest for a Theory of Everything in the third millennium might find itself confronting impassable barriers. We have seen there are limitations imposed by human intellectual capabilities, as well as by the scope of technology. There is no reason why the most fundamental aspects of the laws of nature should be within the grasp of human minds, which evolved for quite different purposes, nor why those laws should have testable consequences at the moderate energies and temperatures that necessarily characterize life-supporting planetary environments. There are further barriers to the questions we may ask of the universe, and the answers that it can provide us with. These are barriers imposed by the nature of knowledge itself, not by human fallibility or technical limitations. As we probe deeper into the intertwined logical structures that underwrite the nature of reality, we can expect to find more of these deep results which limit what can be known. Ultimately, we may even find that their totality characterizes the universe more precisely than the catalogue of those things that we can know.

References

[1] This story seems to have arisen from a misinterpretation of a letter of resignation written by Commissioner Henry L. Ellsworth to his employers, the US Patent Office. The story is scotched by E. Jeffery in "Nothing Left to Invent", *Journal of the Patent Office,* 22, 479–481 (194?).

[2] Barrow, J.D., *Theories of Everything,* Oxford UP, Oxford, 1991. Barrow, J.D., *The World Within the World,* Oxford UP, Oxford, 1988.

[3] Green, M., Schwartz, J., and Witten, E., *Superstrings,* Cambridge UP, 1987. Peat, F.D., *Superstrings and the Search for a Theory of Everything,* Contemporary Books, Chicago, 1988. Green, M. "Superstrings", *Scientific American,* p. 48, September 1986. D. Bailin, "Why Superstrings?", *Contemporary Physics,* 30, 237 (1989). P.C.W. Davies and J.R. Brown, eds., *Superstrings: A Theory of Everything?,* Cambridge UP, 1988.

[4] H.B. Nielsen. *Chaotic Gauge Theories,* World Scientific, Singapore. Barrow, J.D. and Tipler, F.J. *The Anthropic Cosmological Principle,* Oxford UP, Oxford, 1986.

[5] Isham, C., in Russell, R.J., Stoeger, W., and Coyne, G.V., *Physics, Philosophy and Theology,* Univ. Notre Dame Press, IN, 1988. Hawking, S.W., *A Brief History of Time,* Bantam, NY, 1988.

[6] Green and Schwartz, *op cit.*

[7] Barrow, J,D., *The Artful Universe,* Oxford UP, Oxford and NY, 1995.

[8] Barrow, J.D. *Pi in the Sky,* Oxford UP, Oxford, 1992.

[9] Lederman, L. *The God Particle,* Bantam, NY, 1993.

[10] Barrow, J.D., "Observational Limits on the Time-evolution of Extra Spatial Dimensions", *Physical Review D,* 35, 1805 (1987).

[11] Barrow, J.D. *The Origin of the Universe,* Orion, London, 1994.

[12] A. Guth, "The inflationary Universe: a possible solution to the horizon and flatness problems", *Physical Review D.* 23, 347 (1981). J.D. Barrow, "The Inflationary Universe: Modern Developments", *Quarterly Journal of the Royal Astronomical Society,* 29, 101 (1988). Guth, A. and Steinhardt, P., "The Inflationary Universe", *Scientific American,* pp. 116–120, May 1984. Blau, S.K. and Guth, A. in Hawking S.W. and Israel W., *300 Years of Gravitation,* pp. 524–597 Cambridge UP. Cambridge, 1987,

[13] Krauss, L. *The Fifth Essence: The Search for Dark Matter in the Universe,* Basic Books, NY, 1989.

[14] Barrow, J.D. "Unprincipled Cosmology", *Quart. J. Roy. Astron. Soc.,* 34, 117 (1993).

[15] Hartle, J. and Hawking, S.W. "The Wave Function of the Universe", *Physical Review D,* 28, 2960 (1983). S.W. Hawking, *A Brief History of Time,* Bantam, 1988. Halliwell, J. "Quantum Cosmology and the Creation of the Universe", pp. 28–35 *Scientific American,* December 1991.

[16] Penrose, R., *The Emperor's New Mind,* Oxford UP, Oxford, 1989.

[17] Vilenkin, A. "Boundary Conditions in Quantum Cosmology", *Physical Review D,* 33, 3560 (1982).

[18] Chaitin, G. *Algorithmic Information Theory,* Cambridge UP, Cambridge, 1987. G. Chaitin, "Randomness in Arithmetic", *Scientific American,* p. 8O, July 1980.

[19] Hartle, J. and Geroch, R. in *The Quantum and the Cosmos—J.A. Wheeler Festschrift,* W. Zurek, ed., Princeton UP. ??

Chapter 2

THE OUTER LIMITS: IN SEARCH OF THE "UNKNOWABLE" IN SCIENCE

JOHN L. CASTI

I. Logical Barriers in Science

To anyone infected with the idea that the human mind is unlimited in its capacity to answer questions about natural and human affairs, a tour of 20th-century science must be quite a depressing experience. Many of the deepest and most well-chronicled results of science in this century have been statements about what *cannot* be done and what *cannot* be known. Probably the most famous limitative result of this kind is Gödel's Incompleteness Theorem, which tells us that no system of deductive inference is capable of answering all questions about numbers that can be stated using the language of the system. In short, every sufficiently powerful, consistent logical system is incomplete. A few years later, Alan Turing proved an equivalent assertion about computer programs, which states that there is no systematic way of testing a program and its data to say whether or not the program will ever halt when processing that data. More recently, Gregory Chaitin has looked at Gödel's notion of provability from an information-theoretic perspective, finding explicit examples of simple arithmetic propositions whose truth or falsity will never be known by following the deductive rules of any system of logical inference. Essentially, what Chaitin's results show is that such mathematical questions are simply too complex for us.

The theorems of Gödel, Turing and Chaitin are limitations on our ability to know in the world of mathematics. The same limitation applies to statements such as the celebrated Heisenberg Uncertainty Principle in quantum theory, which at first glance appears to refer to an inherent limitation on our ability to measure certain quantities in the physical world. But a more careful examination shows that Heisenberg uncertainty is actually a limitation imposed by certain *mathematical* formulations of quantum theory, and may or may not be a limitation in the real world itself. Similar remarks apply to limitations like Arrow's Paradox in social choice theory and the Central Dogma of Molecular Biology, which are both limitations imposed by *models* of the real world rather than provable limitations about what can be known and/or done in the real world itself. So it's reasonable to wonder if there exist questions about the worlds of natural and human phenomena whose answers science is forever powerless to uncover.

The first, and perhaps most vexing, task in confronting this issue is to settle on what we mean by "scientific" knowledge. Philosophers have grappled with the problem of what constitutes knowledge for ages, with no end to their struggle yet in sight. So to cut through this philosophical Gordian knot, let me advance the perhaps moderately controversial position that the *scientific* way to answer a question makes use of a set of rules, essentially a computer program. We simply feed the question into the rules, turn the crank of logical deduction, and wait for the answer to appear as the output of the program.

But rules come in many flavors, and certainly not all of them qualify as being scientific. For example, the Ten Commandments are certainly a set of rules. Moreover, they help explain the empirical fact that the majority of people are not regularly engaged in robbery, murder or other extreme types of antisocial behavior. But hardly anyone would consider these rules to be in any way "scientific." Similarly, the astrologer's rule "Saturn in the Second House predisposes one to financial misfortunes" is also considered unscientific today, although in an earlier era it might well have been regarded as the height of scientific respectability. This shows, incidentally, that what is and is not scientific is a time-dependent phenomenon, and that scientific rules—or "laws"—are not as absolute as a lot of scientists would like to believe. But if the Ten Commandments and "Saturn in the Second House" are not scientific rules for predicting and/or explaining observed phenomena, what is?

Basically, there are two quite different sets of criteria that must both be satisfied for a rule to have even a chance of being scientific. The first pertains to properties of the rule itself, while the second has to do with the way the rule is arrived at. In regard to the first type of criterion, here is a checklist of characteristics that tend to separate the scientific rules from the pretenders.

- *Explicit*—Scientific rules are explicit, in the sense that there is no ambiguity in the statement of the rule and it requires no private interpretation to employ the rule for prediction or explanation. For example, Newton's laws of motion state an explicit relationship linking the positions, masses and velocities of a collection of material particles. And as long as you understand what is meant by the terms *position, velocity* and *mass,* there is no question about either what the rule says or how it is to be applied.

• *Public*—Scientific rules are open to public scrutiny. They are presented in the open literature and can be tested by anyone who has the time, money, equipment and desire to do so. The contrast here with other types of rules is clear, especially those arising from many religions, where rules accessible only to the "divinely inspired" often play a central role in forming the tenets of a particular system of beliefs.

• *Reliable*—Scientific rules have stood the test of time. Before they are accepted as legitimate laws of nature by the scientific community, the rules must have succeeded in predicting and/or explaining a variety of phenomena over a substantial period of time. Of course, this doesn't mean the rules are infallible and cannot be overthrown in the light of new evidence. But generally speaking, the weight of evidence in favor of the rule must be quite overwhelming before we dignify the rule by labeling it *scientific.*

• *Objective*—Scientific rules are objective in that they are relatively free of investigator bias. In other words, the rule is independent of the social position, financial status or cultural background of the investigator. For example, the exponent in Newton's inverse-square law of gravitation is 2 and not 2.315 or $\sqrt{7}$ or any other number besides 2. And this remains the case for any investigator, regardless of that investigator's professional situation, political leanings or bank balance. In short, the rule is observer-invariant.

I hasten to point out that this does *not* mean that different investigators might not formulate the rule in different terms. But all these formulations must eventually turn out to be equivalent if the rule is to be taken seriously as a scientific rule. For example, in the early days of quantum mechanics there were three seemingly different formulations of quantum phenomena by Heisenberg, Schrödinger and Dirac. Yet upon further investigation it turned out that all three were essentially the same formulation and could be transformed, one to the other, via routine mathematical operations. So what on the surface looked like different rules ended up being the same rule dressed up in different mathematical clothing. This is the kind of objectivity that's characteristic of scientific rules.

So explicitness, public availability, reliability and objectivity are four key properties of rules that tend to separate scientific prescriptions from those that aren't. But earlier we said that these properties constitute just one of two sets of criteria that a rule must satisfy before we label it *scientific*. The other involves the procedure, or method, by which we generate the rule.

If there's even one thing that students in courses on the philosophy of science remember years later, it's the idea of the *scientific method.* This is the process by which many philosophers of science claim that science distinguishes itself from other reality-generation schemes. And it is this process that gives rise to the kinds of rules that serve as candidates for the coveted accolade *scientific.* The principal steps in this process are shown schematically in Figure 1. Here we see three stages: Observation, Hypothesis and Experiment. Traditionally, it's argued that the process starts with Observation, and strictly speaking, I suppose this is indeed always the case. But in a field that's already well developed–particle physics, for example–the diagram may just as often be entered at the Hypothesis stage as at the level of Observation. In any case, after a few tours around the diagram we can hope that the process will converge to something. And that something is what serves as a candidate for a scientific rule. Furthermore, if our use of the scientific method does not converge to something, we generally give up trying to fit the observed phenomena into the framework of science.

Once such filters have been applied to separate scientific rules from the *poseurs,* what remains is pretty much an algorithmic notion of what we usually consider to be a scientific theory. And it is theories we use to answer questions. The problem at this point is that the notion of a set of rules, or an algorithm, is an informal one. Yet to make precise statements about whether or not an algorithm exists for producing the answer to any particular question, we have to formalize the idea of "a set of rules."

Models of Computation

In 1936, Alan Turing published the first truly satisfactory formalization of what we mean by an "algorithm." Turing's idea was what we now call a *Turing machine.* This is a kind of paper computer, consisting of an infinite tape and a scanning head that runs along the squares of the tape, writing/erasing 0s and 1s depending on what it reads on the current square and the "state" (i.e., configuration) the head happens to be in at the current moment. Thus, the Turing machine is specified completely

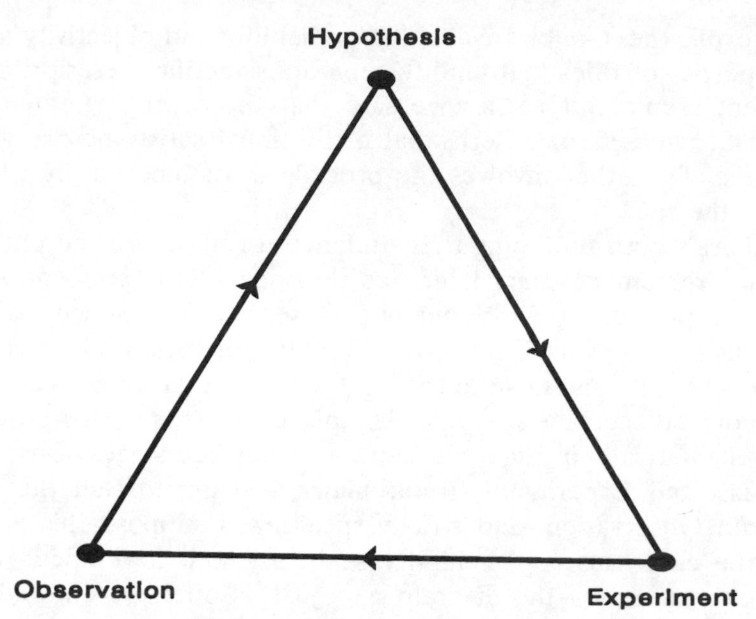

Figure 1. The scientific method.

by its *program,* which is simply a set of instructions telling the scanning head what to do at each moment as a function of what symbol is read on the tape and what the head's current state happens to be. A typical such instruction might be (1, A; R, 0, B), which is interpreted as: "If the tape square currently being scanned contains a 1 and the head is in state A, move Right one square, print a 0 on the new square and enter the state B." Every Turing machine and, hence, Turing machine program, is defined by a finite set of such statements. More details on Turing machines can be found in [1, 8].

The Turing machine is what's called a *model* of computation, since it is a formal mathematical structure defining what we mean by a "computation." But contrary to what some seem to believe, it is not the only model. In fact, a few years ago a very different model was constructed by Blum, Shub and Smale [9]. This BSS model of computation was based on the (informal) idea of a flowchart for a computer program, and gives a different answer than the Turing machine to what is and isn't computable by following a set of rules. So the notion of computability is not an absolute one, but rather hinges upon which model of computation one chooses to employ. The Turing machine model is the de facto standard, and will suffice for the discussion that follows. But it's important to keep in mind that what is scientifically unknowable by the nonexistence of algorithm of the Turing-type may all of a sudden become knowable in the BSS model.

II. The Intractable

Thinking of knowledge, scientific-style, as tantamount to what can be generated by a computer program opens up the issue of computational intractability. We know that there exist puzzles like the celebrated Traveling Salesman Problem whose computational difficulty is widely believed to increase exponentially with the size of the problem. Thus, to calculate the minimal-cost tour of the 180 or so world capital cities by a brute-force examination of each of the 180! possible routings would require a time much greater than the age of the universe with even the fastest of computers. But such a computation is possible—at least, in principle. Let's look just a bit deeper into these types of computational questions.

A famous problem of recreational mathematics is the so-called *Tower of Hanoi.* In this problem, there are three pegs A, B, and C, with N rings of decreasing radii piled on the first peg A. The other two pegs are initially empty. The task is to transfer the rings from A to B, perhaps using peg C in the process. The rules stipulate that the rings are to be moved one at a time, and that a ring can never be placed upon one smaller than itself. Figure 2 shows the problem for the case of $N = 3$ rings.

In this case of three rings, it's not too hard to see that the sequence of seven moves

$$A \to B, \quad A \to C, \quad B \to C, \quad A \to B,$$
$$C \to A, \quad C \to B, \quad A \to B$$

achieves the desired transfer of rings. And, in fact, it can be shown that there is a general algorithm, that is, a program, solving the game for

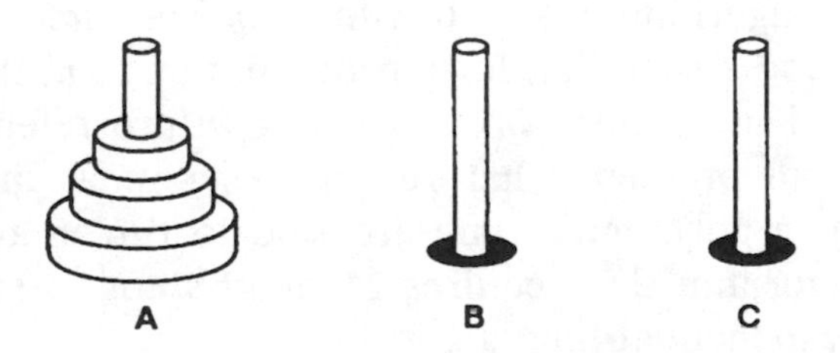

Figure 2. The Tower of Hanoi Game.

any number of rings n. For this program the minimal number of transfers required is $2^n - 1$. Amusingly, the original version of this puzzle, dating back to ancient Tibet, involves $n = 64$ rings. So it's not hard to see why the Tibetan monks who originated the game claim that the world will end when all 64 rings are correctly piled on peg B. To carry out the required $2^{64} - 1$ steps, even performing one ring transfer every 10 seconds, would take well over 5 *trillion* years! Thus, the number of steps needed for the solution of the Tower of Hanoi problem grows exponentially with the number of rings n. This is an example of a "hard" computational problem—one in which the number of computational steps needed to obtain a solution increases exponentially with the "size" of the problem.

By way of contrast, a computationally "easy" problem is the sorting of a deck of playing cards into the four suits in ascending order. First go through the deck until you find the ace of spades. Set it aside and then go through the remaining cards until you find the two of spades, which you also set aside. As one continues in this fashion, the deck is fully sorted. The worst that can happen with this sorting scheme is that the ace of spades is the last card in the unsorted deck, the two of spades is the next-to-last card, and so on. So starting with n cards, you would have to examine at most n^2 cards. Thus, the number of steps needed to completely sort the deck is a quadratic function of the size of the problem, that is, the number of cards in the deck.

In the mid-1960s, J. Edmonds and A. Cobham introduced the idea of classifying the computational difficulty of a problem according to whether or not there exists an algorithm for solving the problem that requires at most a polynomial number of steps in the size of the problem. "Easy" problems can be solved in polynomial time; "hard" problems, on the other hand, require an exponentially increasing number of steps as the problem size grows. The difference in these two rates of growth is shown in Figure 3, where we see that a polynomial function may actually exceed an exponentially increasing one for small values of the size n. But as n continues to grow, the exponential always wins out. To be definite about things, an algorithm is said to run in *polynomial time* if there are fixed integers A and k such that for a problem of size n, the computation will be completed in at most An^k steps. For future reference, we let P be the class of all problems that run in polynomial time. Algorithms that do not run in polynomial time are said to run in *exponential time.* Therefore an algorithm that requires 2^n or $n!$ steps to solve a problem of size n is an exponential-time algorithm.

As an aside, it's worth noting here that this classification into hard

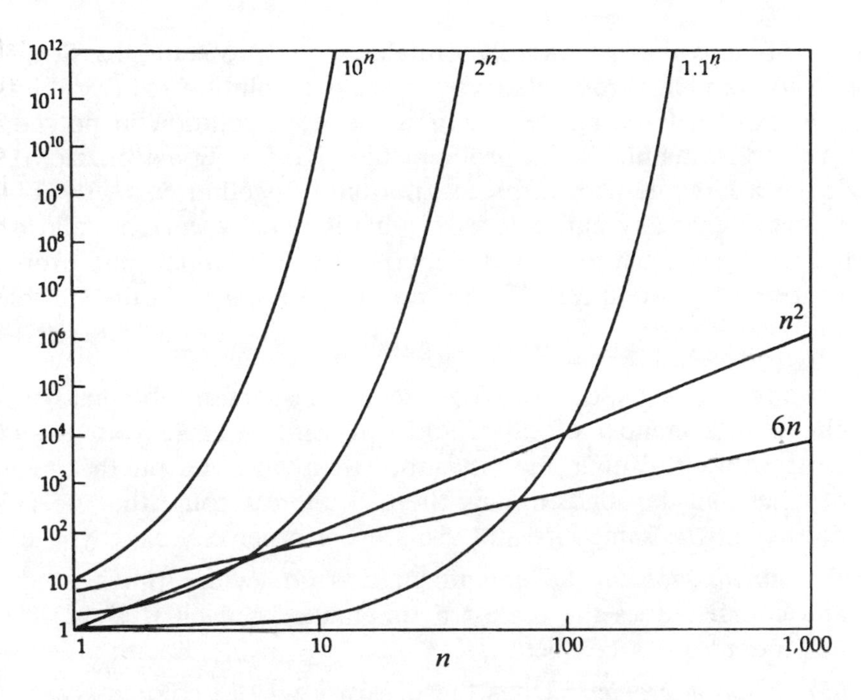

Figure 3. Polynomial and exponential growth of functions.

and easy problems may be a bit misleading when it comes to hands-on computation. For example, an algorithm that has $A = 10^{50}$ and $k = 500$ will still run in polynomial time. But such an algorithm would hardly be "efficient" in any practical sense. On the other hand, an algorithm for the same problem that ran in a number of steps on the order of 2^n might well be preferable to the polynomial-time procedure for small to moderate values of n—even though it is theoretically "inefficient." So the reader should keep in mind that what we are talking about here is the *theory* of computation, not its practice. However, in practice it seems to be the case that problems are solvable either only by exponential-time algorithms or by polynomial-time algorithms that run on the order of something like $10n^2$ or $50n^3$ steps—or less.

P and NP

There is a very important class of problems known as *NP*, which stands for "nondeterministic polynomial time." Let me hasten to point out that this does *not* mean that there is something fuzzy, random, or indeterminate about such problems. Rather, a problem is classified as *NP* if it's possible to verify a proposed solution of the problem in a number of steps that grows polynomially in the problem size. So if you happen to stumble across what you think is a solution to a problem in NP, you can verify or refute that it is indeed a solution in polynomial time. A good example is the problem of solving a jigsaw puzzle. If the puzzle has a large number of pieces, putting it together correctly is hard. But to check that any particular assembly is really a correct solution to the puzzle is easy: just look at it. Clearly, the polynomial-time problems are a subset of those in NP. Here are some examples of other problems that are also in NP:

- *Routing problem:* Suppose you are a salesperson who has to visit clients in a number of cities, and you want to make your round of visits without visiting any city more than once. Given the network of cities and the roads linking them, is there a route that starts and finishes at the same city and visits every other city exactly once?
- *Assignment problem:* Given information on lecture times, students, and courses, does there exist a timetable for each student so that no student has a conflict?
- *Map-coloring problem:* Does there exist a way to color a given map using just three colors, so that no two countries with a common border (greater than a single point) have the same color?
- *Bin-packing problem:* Given a collection of packages, together with their sizes, and a collection of identical bins into which the packages are to be placed, does there exist an assignment of packages to bins such that every package can be placed in a bin without any bin overflowing?

There are many problems in NP for which it's unknown whether or not they are also in P. For example, the question of whether or not n linear constraints on n variables have a solution in integers is of great importance in optimization theory, as we'll see in the next chapter. Yet there is no known polynomial-time procedure for deciding this question. While we have as yet no ironclad proof showing that $P \neq NP$, most computer scientists would be shocked into a state of total catatonia if that turned out not to be the case. One of the main reasons is that a very large number of NP problems have all been shown to be equivalent in the sense that if one of them turns out to be in P, then they are all in P.

This fact, which was proved by Stephen Cook in 1971, is probably the central result of combinatorial computational complexity theory. It serves to motivate work in the field, since it says that in order to refute the $P \neq NP$ assertion, all we need do is produce a single instance of an NP problem for which there cannot exist a polynomial-time algorithm for its solution. But so far no one has managed to find this elusive counterexample. Who knows, maybe the assertion $P \neq NP$ will turn out to be an undecidable proposition, thus independent of the usual axiomatic framework of mathematics, just as the famous Continuum Hypothesis was shown to be neither provable nor unprovable in the 1960s. Perhaps.

But do we really want to wait till the Big Crunch to find the most efficient way to visit the world capitals? And do we really care about the absolutely optimal way to schedule students, courses and classrooms? Wouldn't a quick-and-dirty good approximation serve just as well? It's when we begin asking questions like this about time and money that we reach the point at which science and mathematics part company. Enter the notion of *information-based complexity (IBC)*, in which we try to measure the cost in computational resources to obtain the answer to a question to a predetermined accuracy. Let's look at an example of this sort of unknowability.

Information-Based Complexity

Consider the problem of approximating the definite integral

$$J(f) = \int_0^1 f(t)\,dt.$$

To make things interesting, assume the function f has no closed-form anti-derivative, so that all we have at our disposal with which to compute an approximation to $J(f)$ is n values of the integrand f. Let

$$N(f) = [f(t_1), f(t_2), \dots, f(t_n)]$$

denote the *information pattern* available about the function f. To sprinkle a little analytic structure onto the situation, assume that $f \in C^r[0, 1]$ and that the derivatives of f are uniformly bounded by, say, 1. Thus, the class of functions we consider is

$$\mathcal{F} = \left\{ f \in C^r[0, 1] : |f^{(k)}(t)| \leq 1,\ k = 0, 1, \dots r \right\}.$$

Our goal is to compute an approximation $U(f)$ to $J(f)$ such that

$$|J(f) - U(f)| \leq \epsilon,$$

for all $f \in \mathcal{F}$, where ϵ is some given positive number. Of course, the approximation $U(f)$ is generated by performing various mathematical operations on the information pattern $N(f)$, i.e., $U(f) = \phi(N(f))$, where ϕ is a function defining whatever arithmetic operations, comparison of real numbers, and evaluation of elementary functions needed to compute U for a given information pattern $N(f)$. We define the cost of computing $U(f)$ to be the cost of making the n function evaluations $N(f)$, together with the cost of the operations employed in computing ϕ.

Following in the footsteps of Joseph Traub and Henryk Woźniakowski [6–7], we can now define the computational complexity of the problem of definite integration, $\mathbf{c}(\epsilon)$, to be the minimal cost of computing an ϵ-approximation to $J(f)$ for all $f \in \mathcal{F}$. Note that the quantity $\mathbf{c}(\epsilon)$ is an *intrinsic* property of computing an ϵ-approximation to a definite integral for the function class $\mathcal{F}$. In particular, it does not depend on any specific method. It turns out that for definite integration, $\mathbf{c}(\epsilon) \sim (1/\epsilon)^{1/r}$.

The three pillars underlying IBC are that information is incomplete, noisy and doesn't come for free. So, for instance, in the integration example, the information pattern is the set of function values $N(f)$, together with the knowledge that $f \in \mathcal{F}$. Generally speaking, this is not enough information to pin down uniquely the function f. So the information at hand is only partial. Moreover, the numbers forming the set $N(f)$ are not computed exactly. Usually they come from some kind of numerical operations that contain round-off error or from measurements that are themselves imprecise. So the information is noisy. Finally, information, like everything else worth having in life, costs something to get. Depending on the situation, the cost may come from making the function values or from performing the combinatory operations represented by the function ϕ. But in all cases there is some cost associated with obtaining the information pattern. The primary question studied in IBC is how to obtain an approximate solution to a problem at minimal cost. Let's see how IBC theorists go about addressing this question of basic practical concern in the case of numerical integration.

Recall, the problem is to calculate an ϵ-approximation to the integral

$$J(f) = \int_0^1 f(t)\,dt,$$

where the class of functions is $\mathcal{F}$. As for the algorithm ϕ, one possibility is ordinary Riemann approximation

$$\phi(N(f)) = \frac{1}{n}\sum_{i=1}^{n} f(t_i).$$

In the average-case setting, we seek an approximation $U(f)$ such that

$$\left(\int_{\mathcal{F}} |J(f) - U(f)|^2 \, d\mu(f)\right)^{1/2} \leq \epsilon,$$

where we could choose the probability measure μ to be a truncated Wiener measure on the first r derivatives of f.

In order to define what we mean by computational complexity, we have to have some model of the computational process. A reasonable set of postulates upon which to base such a model are:

1. For every $f \in \mathcal{F}$, the computation of $U(f)$ costs an amount $c > 0$;
2. Each combinatory operation associated with carrying out the algorithm ϕ can be performed without error and costs a unit amount. In other words, we can perform operations on real numbers without error, and each such operation costs us a unit amount.

So if we let **cost** (U, f) denote the total cost of computing the approximation $U(f)$, then we have

$$\textbf{cost}\ (U, f) = \textbf{cost}\ (N, f) + \textbf{cost}\ (\phi, N(f)),$$

where the first term on the right-hand side is the cost of obtaining the information $N(f)$), while the second term measures the cost of combining this information to form the approximation U.

We can now define the computational complexity of a problem to be

$$\mathbf{c}(\epsilon) = \inf\{\textbf{cost}\ (U) : \mathbf{e}(U) \leq \epsilon\},$$

where $\mathbf{e}(U)$ is the error of the approximation U. Note that here we adopt the convention that the cost is infinite if there are no ϵ-approximations. In the two settings we have been discussing, worst-case and average-case, the costs and error functions are given by

– *Worst-Case Setting* –

$$\mathbf{e}(U) = \sup_{f \in \mathcal{F}} \|J(f) - U(f)\|,$$

$$\textbf{cost}\ (U) = \sup_{f \in \mathcal{F}} \textbf{cost}\ (U, f).$$

– *Average-Case Setting* –

$$\mathbf{e}(U) = \left(\int_{\mathcal{F}} \|J(f) - U(f)\|^2 \, d\mu(f) \right)^{1/2},$$

$$\mathbf{cost}\ (U) = \int_{\mathcal{F}} \mathbf{cost}\ (U, f) \, d\mu(f).$$

From these expressions we see that the complexity of a problem involves the intrinsic difficulty in solving that problem, and has nothing to do with the particular algorithm we employ. Rather, it depends on J and $\mathcal{F}$, as well as upon the setting and the set of allowable information construction operations. Finally, it depends on the model of computation we use and, in the average-case setting, on the probability measure μ. So what do these complexities look like for integration?

It turns out that for functions f of smoothness class r, we have

$$\mathbf{c}(\epsilon) \sim \left(\frac{1}{\epsilon} \right)^{1/r}.$$

From this expression, we can get a good handle on the effect of changing the smoothness and/or the error tolerance on the difficulty of calculating the integral of f. For instance, if we go from a once-differentiable function ($r = 1$) to twice-differentiable, then the complexity decreases by a factor of $\epsilon^{1/2}$. It's interesting to note that if we demand only that f be continuous ($r = 0$), then the complexity is infinite! This reflects the well-known fact that continuity doesn't really impose much by way of restrictions on how the function f can wiggle around. And it's the "wiggles" that cause difficulties in determining anything interesting about a function's nature (like its integral) from a finite amount of numerical information. In short, the less structure there is to exploit, the harder will be the computation.

Challenging as such computational problems are, they are not our primary concern here. Rather, our concerns will center on those questions for which there exists *no program at all* for producing an answer. Our interest, then, is in questions that are *logically,* rather than practically, unanswerable. So let me present some specific examples of questions in the realm of natural and human affairs that serve to motivate what's involved in identifying such logical barriers to science.

III. Limits in Nature and in Life

In order to bring the issues of limits to scientific knowledge into sharper focus, let's look at three well-known questions from the areas of physics, biology and economics.

• *Stability of the Solar System:* Certainly the most famous question of classical celestial mechanics is the N-Body Problem, which comes in many forms. One version involves N point masses moving in accordance with Newton's laws of gravitational attraction. Mathematically, the trajectories of the particles are given by the solution of the set of differential equations

$$\begin{aligned} m_j \ddot{\mathbf{r}}_j &= \sum_{i \neq j} \frac{m_i m_j (\mathbf{r}_i - \mathbf{r}_j)}{r_{ij}^3}, \\ &= \frac{\partial U}{\partial \mathbf{r}_j}, \quad i, j = 1, 2, \ldots, N. \end{aligned}$$

Here m_i is the mass and $\mathbf{r}_i$ is the position vector of the ith particle, while $r_{ij} = \|\mathbf{r}_i - \mathbf{r}_j\|$ is the euclidean distance between the ith and jth particles. The quantity

$$U = \sum_{i<j} \frac{m_i m_j}{r_{ij}},$$

is the self-potential (the negative of the potential energy of the particle system).

If we let Δ be the set of all points in R^{3n} where the above set of equations is not defined, then one way of expressing the N-Body Problem is to ask if there is some set of initial positions and velocities of the particles, such that $\mathbf{r}(t) \to \Delta$ as $t \to \infty$? One way for this to happen would be for two particles to collide, in which case $\mathbf{r}_i = \mathbf{r}_j$ for some i and j. Another way would be to have a *noncollision* singularity, in which $\mathbf{r}(t)$ approaches Δ without actually *approaching* any point in this set. In such a case, some particle in the system would acquire an unbounded velocity, and hence "fly off" out of the system. In the special case when $N = 10$, these situations represent a mathematical formulation of the question of whether or not our solar system is stable.

In a 1988 doctoral dissertation, Jeff Xia of Northwestern University used earlier work by Don Saari to give a definitive answer to the second possibility, constructing a 5-body system for which one of the bodies does indeed acquire an unbounded velocity. Figure 4 shows the general

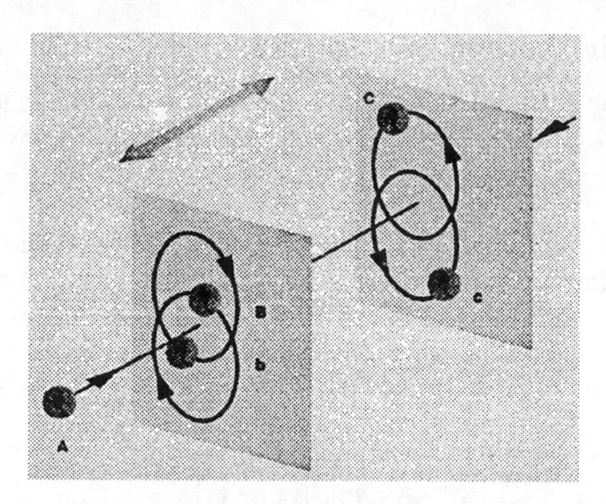

Figure 4. Xia's solution of the *N*-Body Problem.

idea underlying Xia's construction, which involves two binary systems and a fifth body that shuttles back-and-forth between them. By arranging things just right, the single body can be made to move faster and faster between the binaries until at some finite time it acquires as large a velocity as you please. Of course, this result says nothing about the specific case of our solar system. But it does suggest that perhaps the solar system is not stable, and more importantly offers new tools with which to further investigate the matter. For more details on both the history of the *N*-Body Problem and on Xia's work, the survey article [10] is highly recommended.

• *Protein Folding:* The proteins making up every living organism are all formed as sequences of a large number of amino acids, strung out like beads on a necklace. The instructions for what beads go in which positions are contained in the cellular DNA. But once the beads are put in the right sequence, the protein folds up into a very specific three-dimensional structure that determines its specific function in the organism. This process of protein folding is shown in Figures 5 and 6 for *cytochrome-c,* a very short protein consisting of just 104 amino acids, where Figure 5 shows the one-dimensional, linear structure of the amino acids making up the protein, while Figure 6 shows the protein's final, folded-up form.

The big question of protein folding is, given a particular linear sequence of amino acid residues, what three-dimensional configuration will the sequence fold itself into? It is generally thought that the folded

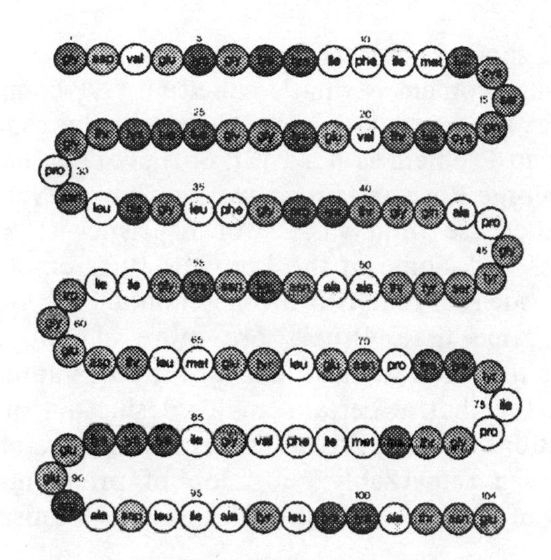

Figure 5. The one-dimensional structure of *cytochrome-c.*

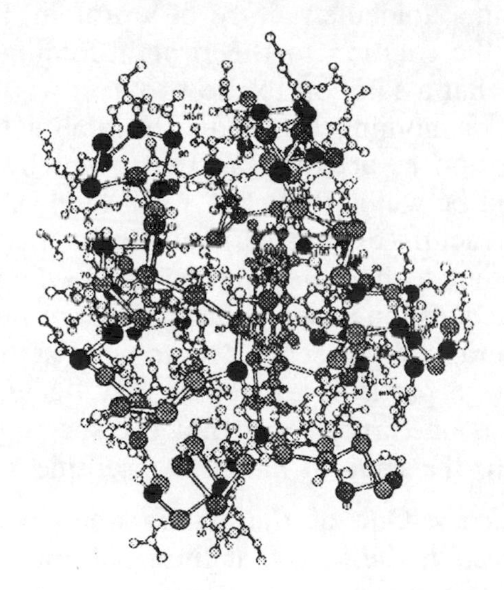

Figure 6. Folded structure of the protein *cytochrome-c.*

configuration of a protein is its lowest free-energy state, and in nature we see proteins composed of several thousand amino acids folding into their final configuration in just a second or so. Yet when we try to simulate this folding process on a computer, it has been estimated that it would take 10^{127} *years* of supercomputer time to find the final folded form for even a very short protein consisting of just 100 amino acids. In fact,

Aviezri Fraenkel showed in 1993 that the mathematical formulation of the protein-folding problem is what's called an "NP-complete" problem, which means that it's computationally "hard" in the same way that the Traveling Salesman Problem is hard [3]. So how does nature do it?

Recently, George Rose and his co-workers have developed a program *LINUS* that makes use of a variety of heuristics to generate folded protein structures [11]. Some of the heuristic rules employed by *LINUS* are basic things, such as that two atoms cannot both occupy the same location at the same time. Others are more chemically specific, for instance, the rule that the chemical facts of life force amino-acid residues to avoid coming together at certain angles. Using a scoring scheme for folded configurations obeying these various rules, Rose et al. have found that *LINUS* does a remarkably good job of predicting much of the folded structure of proteins composed of several thousand amino acid residues.

So here we have a set of rules encapsulated in a computer program for predicting the final, folded structure of a protein. But as Rose points out, *LINUS* is *not* the solution to the protein-folding problem. In fact, it's not even clear what a solution to the problem would look like. Some investigators insist on nothing less than the ability to predict for any protein a conformation as accurate as that given by a high-resolution X-ray. Others would be willing to settle for a kind of bird's-eye view of the basic protein structure as it folds itself up in space.

The general point to be noted here is that when we speak of the inability to answer a question posed about natural or human affairs, there may well be considerable disagreement as to what actually constitutes an answer. So to some, a particular question may be unanswerable, while to others not. This is one crucial way that the sociology and philosophy of science enter into the issue of limits to scientific knowledge.

• *Market Efficiency:* One of the pillars upon which the academic theory of finance rests is the idea that financial markets are "efficient." Roughly speaking, this means that all information affecting the price of a stock or commodity is immediately processed by the market and incorporated into the current price of the security. One consequence of this type of efficiency is that prices should move in a purely random fashion, discounting the natural drift of prices, up or down, due to inflation or deflation. This, in turn, means that trading schemes based on any publicly available information like price histories should be useless; there can be no scheme that performs better than the market as a whole over a significant period of time, or so goes the theory, in any case. For more details on this particular brand of conventional financial wisdom, see [1, Chapter 4].

But actual markets don't seem to pay much attention to academic theory, and the literature is filled with market "anomalies" like the Value Line Theory, the small-firms effect and the low price-earnings ratio effect, all of which cast considerable doubt on the idea of market efficiency [1]. To illustrate this point, the chart in Figure 7 shows a recurrence plot of the Standard & Poor's 500 index over the period 1928–1985. This plot is generated by considering the value of the index at two times t and s, and coloring the point (t, s) in the plane black if the difference in the two values is less than one-half the standard deviation of the entire S&P data set; otherwise, the point remains white. If the market were truly efficient and prices did indeed move randomly, this plot should be shaded a uniform gray; clearly, it isn't. There are bands of white during the Depression, a turbulent time for stocks, and dark black boxes during the late 1950s and early 1960s, which was a period of very low volatility in the market.

So the combination of dozens of market anomalies and pictures like Figure 7 from many different types of financial markets in stocks and commodities strongly suggest that financial markets are not efficient—at least in the strong sense demanded by financial theory. What seems much more likely is that there is a kind of "weak" efficiency in these markets, in which prices fluctuate randomly around the level they would have if strict efficiency were the case. The magnitude of this random

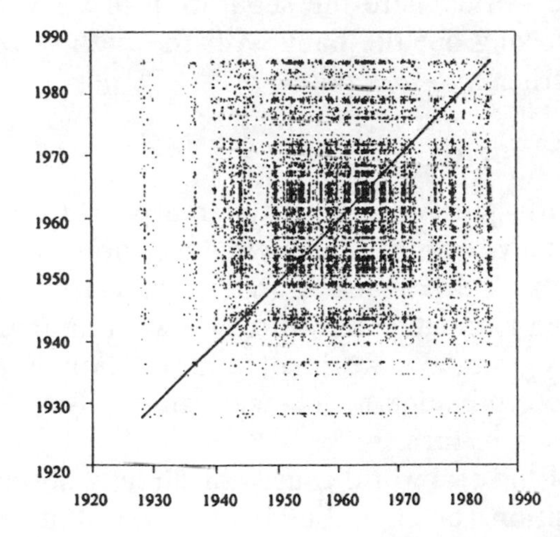

Figure 7. Recurrence plot for the S&P 500 index, 1928–1985.

fluctuation around the efficient price is then a measure of just how inefficient the real market is, and how likely it will then be for some smart market operators to find arbitrage possibilities that would enable them to capitalize on this inefficiency.

So we have seen three major questions about the natural and human worlds—stability of the solar system, protein folding, and market efficiency—and what appear to be three answers: the solar system may not be stable, protein folding is computationally hard and it seems unlikely that financial markets are efficient in any strong sense. But what each of these putative "answers" has in common is that it is the answer to a *mathematical representation* of the real-world question, not an answer to the question itself. So, for instance, Xia's solution of the N-Body Problem says nothing about *real* planetary bodies moving in accordance with real-world gravitational forces. Similarly, Fraenkel's conclusion that protein folding is computationally hard doesn't begin to touch the issue of how *real* proteins manage to calculate the right configurations to do their job in seconds rather than zillions of millennia. And, of course, canny operators on Wall Street have been thumbing their noses at the Efficient Market Hypothesis for decades.

What these examples show is that if we want to look for the scientifically unknowable in the real world, we're going to have to carefully distinguish between the world of natural and human phenomena and mathematical and computational models of those worlds. Unlike conventional scientific investigations in which it usually does no particular harm to mix these worlds, here the separation of worlds is crucial if we want to avoid throwing out the baby with the bath water. It's useful to look just a bit further into this point.

IV. A World of Worlds

Our concern is with questions about the real world that may be unanswerable. Yet the only realm in which we have tools available for *proving* a question to be logically unanswerable is in mathematics. Moreover, since we have taken a scientific answer to be what amounts to the output of a computer program, the world of computation is also directly involved with the determination of unanswerability. Let's quickly examine each of these worlds in turn.

The objects of the real world consist of directly observable quantities like time and position (or quantities like energy that are derived from them). Thus, we consider things like the measured position of planets or the actual observed configuration of a protein. Such observables generally constitute a discrete set of measurements taking their values in some finite set of numbers. Moreover, such measurements are generally uncertain.

In the world of mathematics, on the other hand, we deal with symbolic representations of such real-world observables, where the symbols are often assumed to belong to a continuum in both space and time. Furthermore, the mathematical symbols representing things like position and speed usually have numerical values that belong to number systems such as the integers, the real or the complex numbers, all systems containing an infinite number of elements. And in mathematics the concept of choice for characterizing uncertainty is randomness.

Finally, we have the world of computation which occupies the curious position of having one foot in the real world of physical devices and one foot in the world of abstract mathematical objects. When we think of computation as the execution of a set of rules, an algorithm, then the process is a purely mathematical one belonging to the world of symbolic objects and their relationships one to the other. But if we regard a computation as the process of turning switches ON and OFF in the memory of an actual computing machine, then it is a process firmly rooted in the world of physical observables.

These three worlds—the physical, the mathematical, and the computational—are depicted schematically in Figure 8. And it is the relationship among these very different universes that must be kept uppermost in mind if there is to be any hope of creating a viable *theory* of the limits to scientific knowledge. We must somehow bring these three worlds into congruence insofar as they express any particular question such as, is the solar system stable? Let me close this chapter with a discussion of various approaches to how this kind of congruence might actually be attained.

V. One World or Two?

Let's review the bidding. We are concerned with the existence of questions about the real world that are logically impossible to answer by scientific means. To demonstrate the existence (or nonexistence) of such questions, there are two choices: restrict all discussion and arguments solely to the world of natural phenomena, or mix the worlds of nature and mathematics *cum* computation.

Suppose we choose the first path, remaining exclusively within the confines of the natural world. This means that we are forbidden to translate a question like, "Is the solar system stable?" into a mathematical statement and employ the logical proof mechanism of mathematics to

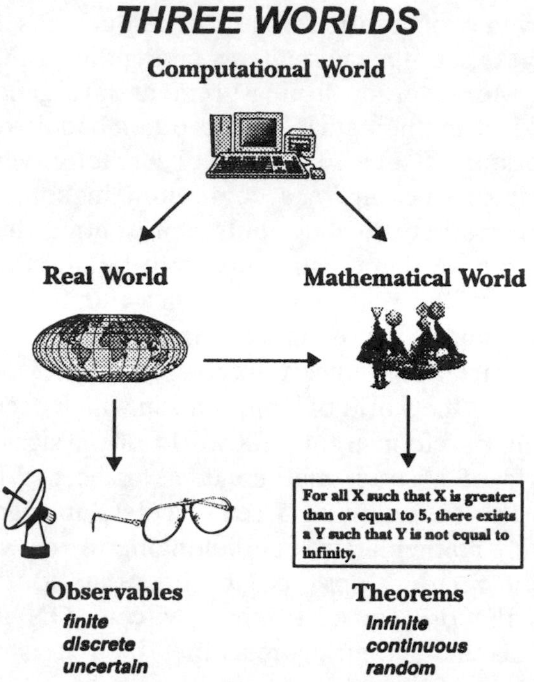

THREE WORLDS

Computational World

Real World **Mathematical World**

For all X such that X is greater than or equal to 5, there exists a Y such that Y is not equal to infinity.

Observables
finite
discrete
uncertain

Theorems
infinite
continuous
random

Figure 8. The worlds of observables, mathematics and computation.

provide an answer. The difficulty here is that our goal is to *prove* that the question is scientifically (un)answerable. But the very notion of a "proof" exists only in the world of mathematics. Therefore, we face the problem of finding a substitute in the physical world for the concept of proof.

A good candidate for replacing "proof" is the notion of "causality." Adopting this as a substitute in the world of natural processes for the mathematical notion of proof, we are led to say that a question is scientifically answerable, in principle, if it's possible to produce a chain of causal arguments whose final link is the answer to the question. So, for instance, the issue of the stability of the solar system might be settled by a causal chain beginning with the positions, velocities, and masses of the planets, which leads via causal arguments to a plot of the planetary orbits. These, in turn, would then show (causally) that no collisions or escapes are possible to within existing observational accuracy. But to construct a convincing causal chain in complicated situations, especially those involving human participants, may be a daunting task. So let's consider approaches that mix the worlds of nature and mathematics.

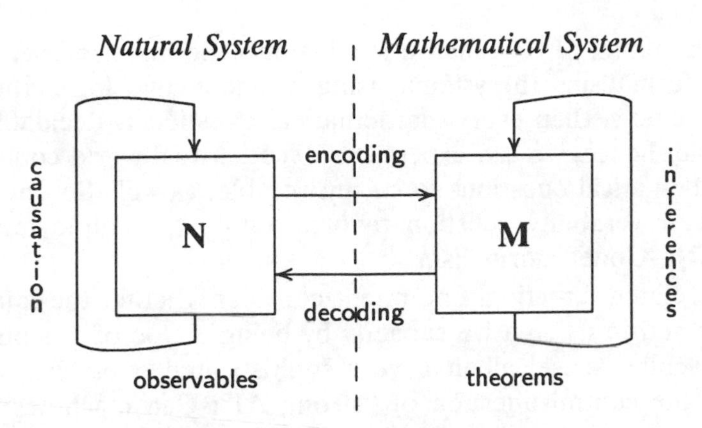

Figure 9. The modeling relation.

If we want to invoke the proof machinery of mathematics to settle a particular real-world question, it's first necessary to "encode" the question as a statement in some mathematical formalism, such as a differential equation, a directed graph or an n-person game. We then settle the *mathematical version* of the question using the tools and techniques of this particular corner of the mathematical world, eventually "decoding" the answer (if there is one!) back into real-world terms. The overall process is shown schematically in Figure 9. This encoding/decoding process sets up what might be termed a "modeling relation" between the real-world phenomena we care about and its mathematical-world representation.

The obvious problem here is the task of providing a convincing argument that the mathematical version of the problem is a *faithful* representation of the question as it arises in the real world. This is simply another way of stating the age-old question of model validity: How do we know that a mathematical model of a natural system and the system itself bear any relationship to each other? This is a *metascientific* question, entailing the development of a theory of models for its resolution. But there is another way.

All existing mathematical results on undecidability rely upon the underlying number system—integers, reals, complex—having an infinite number of elements. But as we noted a moment ago, in the real world of observation measurements must take their value in a *finite* set. There are no Gödel-type undecidability results for logical systems over these kinds of finite number systems. Similarly, employment of nondeductive modes of reasoning—induction and abduction, for instance—take us beyond the realm of Gödelian undecidability. So if we restrict our mathematical formalisms to systems using nondeductive logic and/or finite sets of numbers, then every mathematical question is decidable, that is, answerable; hence, we can expect the decoded real-world counterpart of such mathematical questions to be answerable, as well. So one approach to the unanswerability question revolves about the employment of this kind of "No-Gödel" formalism.

In a related direction, one might consider whether the human mind is constrained in its creative capacity by being a type of computer in the Turing-machine sense, albeit a very sophisticated one. This, of course, is simply the central question of "strong AI": Can machines think just like you and me? Recently, studies of this question under the aegis of the Institute for Future Studies in Stockholm by myself, psychologist Margaret Boden, mathematician Donald Saari, economist Åke Andersson and others suggest strongly that in the creative arts as well as the natural sciences and mathematics, it seems highly unlikely that human creative capacity is subject to the rigid constraints of a Turing-type of computer. Of course, there are other models for computation besides that of Turing [9], and it may well be that the human mind is some type of "super Turing machine," for example, a DNA computer. These matters are currently under investigation. But if the human mind is able to transcend the following of rules in its cognitive activity, then it would also not be subject to Gödelian limits of undecidability.

On this non-Gödelian note, let me conclude this examination of limits to scientific knowledge with a personal speculation about what would be needed for the world of physical phenomena to display the kind of logical undecidability seen in mathematics. Basically, my argument is that for this to happen nature would have to be either inconsistent or incomplete. But what do these mathematical notions correspond to in the world of physical phenomena?

Consistency means that there are no true paradoxes in nature. Quantum mechanics notwithstanding, my view is that particles do not move to the left and to the right simultaneously, nor does water run both uphill and downhill at the same place. Every time we have encountered what appears to be such a paradox, as with the red shift of quasars or the seemingly slow rate of expansion of the early universe, subsequent investigation and theory has provided a resolution of it. So I will take it as an axiom of faith that nature is consistent.

Completeness of nature implies that a physical state cannot arise for no reason whatsoever; in short, there is a cause for every effect. Again,

I can think of no incontrovertible counterexamples to this claim, so I will take it as a working hypothesis that nature is not only consistent, but also complete.

Putting these two assertions—consistency and completeness—together, I believe it's likely that there are no logical barriers to providing a scientific answer to any question we care to put to nature. So perhaps a tour of 20th-century science is not so depressing, after all!

References

[1] Casti, J. *Searching for Certainty.* New York: Morrow, 1991 (paperback edition: Quill Books, New York, 1992).

[2] Casti, J. and J. Traub. "On Limits." Working Paper WP–94–10–056, Santa Fe Institute, Santa Fe, NM, 1994.

[3] Fraenkel, A. "Complexity of Protein Folding." *Bulletin of Mathematical Biology,* 55 (1993), 1199–1210.

[4] Geroch, R. and J. Hartle. "Computability and Physical Theories." *Foundations of Physics,* 16 (1986), 533–550.

[5] Jackson, E. Atlee. "No Provable Limits to Scientific Knowledge." *Complexity,* Vol. 1, No. 2 (1995), 14–17.

[6] Traub, J. "What is Scientifically Knowable?" in *CMU Computer Science,* R. Rashid, ed., Reading, MA: Addison-Wesley, 1991, pp. 489–503.

[7] Traub, J. and H. Woźniakowski. "Breaking Intractability." *Scientific American,* 270 (January 1994), 102–107.

[8] Davis, M. *Computability and Unsolvability.* New York: McGraw-Hill, 1958 (Dover reprint, 1982).

[9] Blum, L., M. Shub and S. Smale. "On a Theory of Computation and Complexity Over the Real Numbers: *NP* Completeness, Recursive Functions, and Universal Machines." *Bull. Amer. Math. Soc.,* 21 (1989), 1–46.

[10] Saari, D. and Z. Xia. "Off to Infinity in Finite Time." *Notices Amer. Math. Soc.,* Vol. 42, No. 5 (May 1995), 538–546.

[11] Rose, G. "No Assembly Required." *The Sciences,* January-February 1996, 26–31.

Chapter 3

VARIATIONS ON AN ORIGINAL THEME

N. C. A. da Costa and F. A. Doria

Dedicated to the memory of
Professor Lucia Ferreira Reis

Johannes Brahms and Gustav Mahler became close friends when the younger composer was appointed Director of the Vienna Opera. René Leibowitz [20] *tells that Brahms and Mahler once walked through the woods near Vienna while chatting about the future of music. Brahms's views were that he belonged to the last wave in the great musical movement that had begun with Bach and went on with Mozart, Beethoven and Schubert. When they happened to pass a brooklet, Mahler retorted, "There goes the last wave."*

Science is part of our historical and cultural heritage. We can trace its seeds to Parmenides, as Heidegger does, when out of the blue, in the midst of a comment on a tricky verse in Parmenides' Perì phúseos, *he starts to relate his analysis of Pre-Socratic thought to nuclear plants* ([16], p. 142),

> Wäre das Sein des Seienden nicht als Anwesen des Anwesenden offenbar, dann hätte niemals die elektrische Atomenergie zum Vorschein kommen ...

and computers ([16], p. 145),

> Die abendländische Logik wird schließlich zur Logistik, deren unaufhaltsame Entfaltung inzwischen das Elektronenhirn zeitigt ...

Or we can find the roots of modern science in the mystico-religious environment of John Dee and Johann Valentin Andreæ in the late 16th and early 17th centuries, as Frances Yates suggests [33], *when carefully tracing back to Dee and Andreæ the origins of the Royal Society. Or we can even take a more cautious, conservative outlook on the beginnings of science.*

Never mind. As part of our historical process, science is most certainly doomed to fade away. There one may find the ultimate limits to our present scientific endeavor.

Yet, we must pay attention to Mahler's remark.

I. *Leitmotiv:* The Halting Problem is Solvable in the Language of Analysis

Let's step into a time machine and go back to the fifties, just when the first commercial computers were being developed. Let's enter a computer lab. A technician is now giving a set of instructions to the machine. The machine starts its number-crunching procedures, and goes on and on; it never stops. The machine has entered an infinite loop. The technician complains about the need to develop a program that would check for those infinite loops; out of the input and out of the program coded into the machine's belly, one would know beforehand whether it stops or not. "No one has so far developed one such program," the technician tells us. No surprise, since the technician asks for an algorithmic solution of the algorithmically unsolvable Halting Problem.

The Halting Problem is older than computer science itself, and was proved to be algorithmically unsolvable by Turing in his 1936 paper on the *Entscheidungsproblem* (see [21] and [25] for details and references). The Halting Problem can be fully discussed within intuitive mathematics, and gives us the backbone for one of the arguments that lead to the incompleteness phenomenon within formalized arithmetic and its extensions.

Suppose that we have embedded Turing machine theory inside Zermelo-Fraenkel set theory, which from now on we will call T. It is strong enough to prove all intuitively derivable results about those machines. Then there is a Turing machine M and an input value a such that the assertion "$\mathsf{M}(a)$ never stops" is unprovable from the axioms of T, as any mathematical proof is an algorithmic procedure, in the usual sense for 'mathematical proof' (but see Section 5 below). The provability of all statements like "$\mathsf{M}(a)$ never stops" would then imply that we can solve the Halting Problem.

All our results here stem from this example [6], since the Halting Function $\theta(m, n)$ is our unifying, explicit motive. Everything here follows from it. (Note: The Halting Function $\theta(m, n)$ is a function that solves the

Halting Problem; it cannot be algorithmically expressed. Yet there are explicit expressions for the Halting Function $\theta(m, n)$ in formal languages quite close to arithmetic.)

The Halting Function

Lemma 1.1. *The Turing machine $M_m(n)$ of index m halts over n if and only if $\theta(m, n) = 1$; $M_m(n)$ doesn't halt over n if and only if $\theta(m, n) = 0$, where θ is given by*

$$K(m, n) = \int_{\mathbb{R}^q} c(m, n, x_1, \ldots, x_q) \exp\left[-((x_1)^2 + \ldots + (x_q)^2)\right] d\Omega,$$
$$\theta(m, n) = \sigma\left[K(m, n)\right]. \square$$

The function θ is the Halting Function of computer science [21, 25]. This Lemma shows that an expression for it exists within classical analysis. For the proof see [6]. The quantity c is obtained from Richardson's transforms (see the reference), while σ is the sign function: $\sigma(\pm x) = \pm 1, \sigma(0) = 0$.

A still simpler expression for the Halting Function is available within languages quite close to arithmetic [9]; one only has to be able to make infinite sums in order to obtain it. Let $p(n, \mathbf{x})$ be a 1-parameter universal polynomial, where $\mathbf{x} = (x_1, \ldots, x_p)$. Then either $p^2(n, \mathbf{x}) \geq 1$ for all $\mathbf{x} \in \omega^p$, or there are some $\mathbf{x}$ in ω^p such that $p^2(n, \mathbf{x}) = 0$. Note that here ω is the set of positive integers. Since $\sigma(x)$ is primitive recursive when restricted to ω, we may define a function $\psi(n, \mathbf{x}) = 1 - \sigma p^2(n, \mathbf{x})$ such that

- Either $\psi(n, \mathbf{x}) = 0$ for all $\mathbf{x} \in \omega^p$, or
- there exist $\mathbf{x} \in \omega^p$ so that $\psi(n, \mathbf{x}) = 1$.

Thus the Halting Function can be represented as:

$$\theta(n) = \sigma\left[\sum_{\tau^q(\mathbf{x})} \frac{\psi(n, \mathbf{x})}{\tau^q(\mathbf{x})!}\right],$$

where $\tau^q(\mathbf{x})$ denotes the positive integer obtained from $\mathbf{x}$ by the pairing function τ^q, which maps q-tuples of positive integers onto a positive integer, and is such that $\tau^{q+1} = \tau(x, \tau^q(\mathbf{x}))$.

Moreover, given any family of Diophantine equations $p(n, \mathbf{x}) = 0$ parametrized by $n \in \omega$, we can obtain a halting-like function $\theta_p(n)$ that equals 1 whenever $p(n, \ldots) = 0$ has integer roots, and equals 0 if there are no such roots. The procedure for the construction of θ_p goes in the same way as in the construction of the Halting Function $\theta(m, n)$.

II. First Variation: From Rice's Theorem to the Undecidability of Classical Analysis

A recent survey of problems encountered in the development of large software systems [15, 31] considers the difficulties of predicting the behavior of those monster computer programs. Practical (and costly) consequences are delays in their release—delays seen, for example, in the release of operating system *Windows 95* and the opening of Denver's new airport, which was "postponed to allow [its software developers] time to flush the gremlins out of its $193 million system." It has been suggested that the trouble may lie in the fact that software engineering is a new discipline, akin to chemical engineering in its first stages, before some kind of unifying theory was developed for it. (In the case of chemical engineering, the theory underlying it involves the so-called "unit operations.") According to that view, software engineering might still be looking for its unifying theoretical background.

But let's take a closer look at the difficulties encountered in the development of those new, large-scale, computer programs. They involve

• *"Gremlins"*—these are bugs and erratic behavior of the software in not-so-unusual situations, and

• *Delays*—the slowness of the program's response to normal, everyday inputs.

The latter difficulty points to the main issues in computational complexity theory (exponential behavior, for instance, is immediately perceived in image-rendering programs such as *Bryce*); the former problem revolves about Rice's Theorem, a classical result in recursion theory.

Rice's Theorem can be paraphrased simply: There is no decision procedure to test for properties of the output of an arbitrary program (trivial cases excluded). So, debugging will always be a tentative, basically empirical, procedure. Let's be just a bit more specific. (The notation, conceptual starting points and proofs underpinning what follows can be found in [9, 25, 27].)

We say that C is an *index set of algorithms* if, given an algorithm $\phi \in C$ and another algorithm ψ, with $\phi(n) = \psi(n)$ for all positive integers n in both domains, then $\psi \in C$. An index set C is *trivial* if either $C = \{\emptyset\}$ or C is the set of all programs. Then we have

Rice's Theorem. *C is recursive if and only if it is trivial.*

The halting function implies a general undecidability and incompleteness theorem.

In 1990, the authors proved a theorem similar to Rice's Theorem within the language of classical analysis [9]. Given a formal version of the language of advanced calculus, where we only handle elementary functions and calculus operations (that is, polynomials, sines and cosines, special constants such as π and e, the absolute value function $|\cdots|$, derivatives and integrations over simple domains in $\mathbb{R}^n$), we established a result that strikingly parallels Rice's Theorem. (That result immediately leads to a surprisingly "user-friendly" technique to prove undecidability and incompleteness theorems in the formal counterpart of classical mathematics.)

We take ZF (Zermelo–Fraenkel set theory, supposed consistent) as our formal theory, plus some portions of the Axiom of Choice. We also require the ι symbol to make room for definitions, so that in fact we deal with a conservative extension of the language of ZF. That formal theory and language, call it T, is enough for our purposes.

Now let P be a predicate in the language for classical analysis (intuitively, a predicate is any assertion about the objects of that language). We say that P is *trivial* if and only if it provably applies to either all or none of the terms in that language. Then:

Theorem (da Costa and Doria). *P is recursive if and only if P is trivial.*

(We say that P is recursive whenever we have a decision procedure for it.)

So, there are no decision procedures for nontrivial properties. We can't even decide whether an arbitrary expression in the language of analysis equals 0. (Consider, for instance, the Halting Function θ.)

For example, think of P as the assertion "x is even," with the domain of P being the set of integers. Then there is no general algorithm to check whether an arbitrary expression ξ is an even integer.

We don't need the axiomatic background in order to prove the preceding *undecidability* result; it will only be required in order to prove the related *incompleteness* result stated below.

In order to obtain the incompleteness version of that general undecidability result, we need a lemma derivable from those results. Recall that "$T \vdash \alpha$" means that α can be derived from the axioms of T. Then we have:

Theorem. *There is an explicit expression for a function β such that $T \vdash \beta = 0$ or $\beta = 1$, while neither $T \vdash \beta = 0$ nor $T \vdash \beta = 1$.*

(In case you are curious, the proof follows from the fact that the θ function has a recursively countable set of nonzero values, with a nonrecursive complement, together with the existence of a Diophantine equation having no roots in a model **M** with a standard arithmetic portion that can neither be proved nor disproved by T.)

Then we have the result,

Theorem. *For every nontrivial P, there is an expression x in the language of T such that neither $T \vdash P(x)$ nor $T \vdash \neg P(x)$.*

(If $P(x_0)$ and $\neg P(x_1)$ are both provable, write $x = x_0\beta + x_1(1 - \beta)$.)

An Application: There is No General Algorithm to Check Whether a Real Number is Transcendental

Our techniques lead to several interesting results in both pure and applied mathematics. An old question in number theory deals with the existence of algorithms to check for the transcendence of a real number: Is there a mathematical recipe for it? We know for sure that there are proofs of the transcendence of individual real numbers (e and π, for example) and several sufficiency characterizations, such as Liouville's Theorem [23].

Irrational numbers are usually exhibited with the help of series expansions, integrals or by a combination of both; arguments frequently include elementary functions and the absolute value function $|\cdots|$ (whenever, for instance, we require absolutely summable series). Consider the set of all real numbers that can be written in this fashion. Then:

> There is no general algorithm to check whether one of those otherwise arbitrary expressions for a real number represents a transcendental number.

(If r_0 is a provably algebraic number and if r_1 is provably transcendental, then the family of numbers $r(m) = r_0\theta(m) + r_1(1 - \theta(m))$ cannot be checked for transcendentality.)

Another result along these same lines is:

> If real numbers are as given above, then there is no general algorithm to decide whether or not a real number is computable.

(Recall that a real number is computable if its binary expansion is the image of a total recursive function [25]. The preceding result shows that we may expect trouble when checking for computability in complicated expressions.)

Rice's Theorem in the Language of Analysis

The relation between Rice's theorem and the authors' undecidability results isn't a simple one, however. In order to clarify it we can proceed as follows.

1. Code programs (given by Turing machines $M_m(n)$) as parametric families of Diophantine equations $p(\langle m, n\rangle \ldots) = 0$, where the polynomial equation $p\langle m, n\rangle = 0$ has integer solutions if and only if $M_m(n)$ stops on input n.
2. Use Richardson's map ρ [6] to code those expressions into expressions in the language of analysis.
3. Given a set of programs C, we generate a set ρC of expressions in the language of analysis.
4. There is then a decision procedure for $\xi \in \rho C$ if and only if there is one for $\rho^{-1}\xi \in C$, since the above described handling of programs, Diophantine equations and expressions in analysis is fully algorithmic.

Step (4) results in a translation of Rice's Theorem into the language of analysis. It implies that the language of analysis is undecidable, since there is a set of properties $\{\rho C\}$ of expressions in analysis that is undecidable (from Rice's result). However, our own results go beyond that, showing that *every* nontrivial predicate P in that language is undecidable and leads to undecidable sentences, and not just a strict subset $\{\rho C\} \subset \{P\}$.

III. Second Variation: Higher Arithmetic Degrees in Analysis

The language of classical analysis is very rich. We can obtain infinitely many expressions for the halting function within it, and therefore we can enter those expressions as oracles into Turing machines and obtain explicit expressions for characteristic functions of higher arithmetic degrees in the same language [9]. Our main result asserts (roughly) that we can obtain explicit expressions for functions, such as that their computation is as difficult as one wishes, up to the arithmetical hierarchy—and beyond.

Again, everything here stems from the halting function θ. The first degrees beyond arithmetic can be dealt with in classical mathematics. But as problems related to the continuum hypothesis or to the axiom of choice begin to set in as one goes far enough, the mathematical situation becomes richer, the farther one goes from the arithmetical hierarchy. (The point is that we have to deal with different standard models for Zermelo-Fraenkel set theory when one reaches beyond arithmetic.)

More Mathematical Results

Proposition 3.1.

1. *Given any nontrivial predicate P in T, there is an infinite denumerable family of expressions ξ_m, such that for those m with $T \vdash$ "$P(\xi_m)$," for an arbitrary total recursive function $g\colon \omega \to \omega$, there is an infinite number of values for m such that the shortest length $C_T(P\xi_m)$ of a proof of $P\xi_m$ in T satisfies $C_T(P\xi_m) > g(\|P\xi_m\|)$. (Note: Recall that ω is the set of positive integers.)*
2. *Given any nontrivial predicate P in T, there is an expression ξ such that $T \vdash$ "$P(\xi)$" if and only if $T \vdash$ ("Fermat's Conjecture").*
3. *There is an expression ξ such that for a nontrivial P, $P(\xi)$ is T-arithmetically expressible as a Π_{m+1} problem, but not as any Σ_k problem, $k \le m$.*
4. *There is an expression ξ, such that $P(\xi)$ is not arithmetically expressible.* □

IV. Third Variation: Faceless Objects

We can also obtain in that same language of T explicit expressions for "faceless" objects.[1] A "faceless" object is a set in T all of whose

[1] F. Miró–Quesada, who gave the name "paraconsistent logics" to the logical systems developed by N. da Costa, has recently suggested that those 'faceless objects' be called *afacial* in the main Latin languages.

"interesting" or "meaningful" properties are formally undecidable. Those objects remind us of Cohen's set-theoretic generic subsets of the integers [5], and in fact there is a coding from our faceless objects to a Cohen-generic subset of the integers that is a strict subset of Cohen's original example [10]. So we place those rather unexpected results of ours solidly within the framework of classical set theory.

A (fully) faceless object in a ZF-set X is an element of the set X, such that all its nontrivial properties are undecidable within the formal theory (besides that of being a member of X and similar noninformative properties). That object has been explicitly represented by the authors as an infinite expansion, very much like Hilbert space expansions in quantum mechanics. (The similarity to objects in quantum mechanics and in quantum field theory is uncanny, as the undecidable coefficients are formal look-alikes of Feynman integral expansions. So far, however, we haven't explored that similarity.)

We start with a T-set X of objects defined by some predicate Q, say, $X = \{x \in y : Q(x)\}$. Let $P_i, i \in \omega$, be an enumeration of all predicates in the language of T such that there are $z_{0i}, z_{1i} \in X, z_{0i} \neq z_{1i}$ with $T \vdash P_i(z_{0i})$ and $T \vdash \neg P_i(z_{1i})$. 'Components' of the faceless object are precisely the representatives z_{0i}, z_{1i}, of all properties that affect the elements of X, while the coefficients that multiply those 'components' are undecidable expressions whose (undecidable) values are either 0 or 1. We can show in T that only a single coefficient will survive, while the rest will vanish, but there is no decision procedure to check which coefficient is the nonvanishing one.

These faceless objects remind us immediately of Cohen's generic sets, which have undetermined properties fixed by the forcing conditions. To be more specific, in our construction the quantum-mechanical-like undecidable coefficients stand out as natural codings for the Boolean values assumed by a Cohen-generic set in Boolean-valued set theory. See [10] for details.

We can also obtain expressions with controlled properties, say, such that only a finite number of those predicates for them is decidable. Let $P_i, \ldots, P_k$ be a finite listing of nontrivial predicates that provably apply to some Q-objects (and not to others). Take the conjunction $\bigwedge_i P_i$ as a more restricted condition Q' and proceed as above.

'Faceless' Objects

Our construction of the 'faceless' object proceeds as follows. Let the quantity $Q(x)$ be a T predicate. Think of that predicate as, say, the characterization of the set of all hamiltonians on all symplectic manifolds. An object that satisfies Q is a *Q-set,* which here is simply an arbitrary hamiltonian.

Suppose we are given an enumeration of all the predicates P_k in the formal theory T. Moreover, we suppose that for some z_i, $T \vdash Q(z_i)$, $i \in \omega$. Out of all those formal expressions z_i that provably satisfy Q, we pick those pairs z_i, z_j, such that the P_k can be proved to be satisfied by a member of the pair and provably denied by the other member of the pair.

Out of that we can obtain in T a recursively countable set of pairs of expressions x_{2i}, x_{2i+1}, i a natural number, that represent different objects and such that:

1. Both $T \vdash Q(x_{2i})$ and $T \vdash Q(x_{2i+1})$.
2. $T \vdash P_i(x_{2i})$ and $T \vdash \neg P_i(x_{2i+1})$,

where the P_i are nontrivial predicates (relative to Q) in T that range over Q-sets.

In other words, we list everything that provably applies to the Q-sets and pick a representative for each property in the listing, as well as for its complement. Whenever z is a 'faceless' object, the theory becomes blind to the particular details—there are none to be checked. So we can only prove that z is a Q-object and nothing more.

The intuition behind set-theoretic generic objects is similar; set-theoretic generic objects are also faceless, having no properties beyond those that are 'forced' upon them by the forcing conditions. So, can we go from z to a Cohen-generic set? Definitely [10]! Then:

Proposition 4.1. *Within T:*

- **(Undecidability)** *There is a countable family z^*_m of expressions for Q-sets, such that there is no general algorithm to decide for any nontrivial Q-predicate P_k in T whether one of the z^*_m satisfies (or doesn't satisfy) P_k.*

- **(Incompleteness)** *There is an expression for a Q-set z all of whose nontrivial Q-properties cannot be proved within T.* □

Let ξ_1, ξ_2, $\xi_3, \ldots$ be an infinite countable sequence of mutually independent Gödel sentences in T. Let $m_1, m_2, \ldots$ be the corresponding Gödel numbers. Form $\theta_1 = \theta(m_1), \theta_2 = \theta(m_2), \ldots$ and define $\epsilon^1_j =$

$1-\theta_j, \epsilon_j^0 = \theta_j$, for all $j \in \omega$. Let τ_n be a variable that ranges over all 2^n binary sequences of length n, coded by ordered n-tuples of zeros and ones. Finally, establish a map f between those n-bit binary sequences and all n-factor products $\epsilon_1^{\alpha}\epsilon_2^{\alpha'}\dots\epsilon_n^{\alpha^n}$, so that in the i-th position $0(1) \mapsto \epsilon_i^0(\epsilon_i^1)$. Given τ_n, the associated product is $f(\tau_n)$. Note that given a specific model for T, all such products equal 0, save a single one, which equals 1. The actual nonzero product depends on which model is chosen.

We now order the predicates $P_1, P_2, \dots$ in some fixed way. We will assume that $T \vdash P_i(z_{2i-1})$, while $T \vdash \neg P_i(z_{2i})$. Next, list all finite binary sequences and select from those a countably infinite set of mutually incompatible sequences (incompatible sequences are those that do not have common extensions) of growing length (such as, say, $0, 00, 000, \dots, 1, 01, 001, 0001, \dots$). Given each binary sequence, it is either in our previously chosen set of incompatible sequences or will be the extension of some previously chosen sequence. Note the incompatible set $\{\tau_1, \tau_2, \dots\}$; if τ_j is any sequence, $\tau_{j'}, \tau_{j''}, \dots$ is its extensions. (A general extension is denoted τ_{j^k}.) The expression we require is

$$z = z_1[(1/2)f(\tau_1) + (1/2)^2 \sum{}^* f(\tau_{1'}) + \dots]+$$
$$z_2[(1/2)f(\tau_2) + (1/2)^2 \sum{}^* f(\tau_{2'}) + \dots] + \dots \quad (1)$$

(The $\sum^*$ is a sum over all extensions of equal length in the ϵ.)

It remains to show that we have $T \vdash$ "z is an expression for a Q-set," while none of the remaining properties P_i for Q-sets can be verified. The last statement is immediate from the construction; the first statement follows from the next discussion, which is split into two parts. Let us be given the B_j:

$$B_j = \left[(1/2)f(\tau_j) + (1/2)^2 \sum{}^* f(\tau_{j'}) + \dots\right],$$

where the $f(\tau_j)$ are (the above given) products of undecidable zero/one functions coded by initial segments of binary sequences, such that $B_j \perp B_k$, $j \neq k$; the $\tau_{j'}$ are 1-extensions of the τ_j, and the sum $\sum^*$ is over extensions of equal length and so on. The point is that the B_j are (formal) expansions of undecidable functions. Yet we can assert

Lemma 4.2. *$T \vdash$. "There is a single $i_0 \in \omega$ such that $B_{i_0} = 1$, while for all $i \in \omega$, if $i \neq i_0$, then the $B_i = 0$."* □

Corollary 4.3. *$T \vdash Q(z)$, where $z = \sum_j B_j z_j$, the z_j being as above.*

PROOF: Since a single $B_{j_0} = 1$, z always equals some Q-set. □

(The first assertion of Proposition 4.1 is immediate. See the details in [8, 9, 11].)

From the Halting Function to Set-Theoretic Genericity

We summarize the main ideas. Let **V** be a model for ZF plus the axiom of choice (ZFC), and let $B = \mathrm{RO}(2^\omega)$ be the regular open algebra of 2^ω [2]. Form the Boolean extension $\mathbf{V}^B$. Now make

Definition 4.4. $B_k = \{g \in 2^\omega : \tau_k \subset g\}$.

Then we have that $u(\hat{n}) = B_n$, for $n \in \omega$. Notice that the coefficients in Eq. (1) are a coding for the B_k; for instance:

$$\left[(1/2)f(\tau_1) + (1/2)^2 \sum{}^* f(\tau_{1'}) + \ldots\right] \mapsto B_1.$$

Those coefficients are therefore (in a natural sense) a coding for the Cohen 'name' of the generic set in **V**.

Now let v be the usual example for a Cohen-generic set in a Boolean-valued universe [2]. Then:

Proposition 4.3.

1. $\mathbf{V}^B \models$ *"u is set–theoretically generic."*
2. $\mathbf{V}^B \models$ *"$u \subset v$ and inclusion is proper."* □

V. Fourth Variation: Beyond Turing Computation

Let's go back to the original intuitions on the finitary aspects of computability, as described by Hilbert himself:

> *Wir [wollen] allemal mit dem Wort* finit *zum Ausdruck bringen, daß die betreffende Überlegung, Behauptung oder Definition sich an die Grenzen der grundsätzlichen Ausführbarkeit von Prozessen hält und sich somit im Rahmen konkreter Betrachtung vollzieht* ([17, p. 32]).

The point here is the meaning of *finit,* "finitary." According to Gödel, finitary is everything that can somehow be reduced to our 'concrete intuition':

> *"Concrete intuition," "concretely intuitive" are used as translations of* Anschauung, Anschaulich. *The simple terms "concrete" or "intuitive" are also used in this sense .... What Hilbert means by* Anschauung *is substantially Kant's space-time intuition confined, however, to configurations of a finite number of discrete objects* ([14, p. 272, note b]).

We propose to go beyond Hilbert's characterization and to introduce some kind of *analog* reasoning, as that may shorten difficult proofs. Here are some examples.

• *The Jordan Curve Theorem.* Intuitively, this result says that given a circle C on the plane R^2, the circle divides R^2 into two disconnected regions: the set of points inside C and the set outside. A rigorous statement is only slightly more complicated.

The visual proof is immediate; one simply draws the corresponding picture and *sees* it. In a word, it is visually *intuitive.* However, a rigorous proof is notoriously complex [see 22, pp. 110-120, for instance].

• *Finite Covers for a Compact Planar Domain.* A disc D in the plane $\mathbb{R}^2$ is a compact domain. Therefore, any of its infinite covers has a finite subcover. Now cover D with an arbitrary number of open discs, and look for a subcover of that cover with the following properties:

1. The closure of each disc from the cover lies in the interior of D.
2. Those closures do not intersect each other.

So, we fill D with smaller discs that don't touch each other and do not touch the inner side of the boundary of D. Is there a *finite* subcover of this type? Obviously not—draw the picture and ponder it for a moment. But again a rigorous proof is exceedingly difficult [12].

• *The Shortest Path Between Two Arbitrary Cities.* Pick two cities on a map, say, San Francisco and New York, and try to find the shortest road linking them. The computational solution of this problem is very hard. But we can proceed in an analog fashion as follows: Draw the mesh of roads between San Francisco and New York with strings and pull at the endpoints representing the two cities. The tight string(s) represents the shortest route.

How is it that a rigorous mathematical proof for those intuitively obvious facts is so complicated? In a nutshell, the reason is that mathematics restricts its proof tools to *digital,* discrete, stepwise techniques, while intuition makes use of *analog-like* arguments. The point is that something seems to be missing in our current formalizations of the idea of "mathematical proof," since they cannot reproduce the immediateness of our geometrical intuition.

A Super-Turing Computation Theory

We sketch here the theoretical basis for a computation theory (H-computation theory) that is strictly stronger than the Turing-Church model. An algorithm in that extended theory obeys the following (non-rigorous, intuitive) deterministic criteria:

1. An algorithm is a set of instructions given by a finite sequence of discrete symbols.
2. The input for that algorithm is also a finite sequence of discrete symbols.
3. The actual computation takes place in discrete steps, finite in number, so that any computation can be coded as a finite sequence of discrete symbols.
4. If and when the computation terminates, its output is expressed as a finite sequence of discrete symbols.
5. *(Geometrical Principle.)* We can always decide whether two smooth lines within a rectangle in the plane intersect.

The Geometrical Principle was suggested by the first two examples given above. It is enough to extend our theory strictly beyond Turing computation. (Actually we won't even require its full strength.)

Those ideas were first suggested in remarks scattered throughout some of our papers, such as the final comments of [6] and [7]. For another analog-like, super-Turing computation theory see [26]. The main result is:

Arithmetic is H-decidable.

Standard and Nonstandard Models and H-Computation; a Super-Halting Problem

We have indicated elsewhere [9, 19] that a Gödel-undecidable sentence with respect to an axiomatic system such as ZFC may be viewed as a bifurcation point in a representation of that axiomatic theory through some convenient dynamical system. (For one such coding of formal systems by dynamical systems, see [9].) H-computation decides arithmetic theories along the *standard* model, while the bifurcation-point picture for undecidable sentences suggests some sort of equal weight for truth decided in either standard or nonstandard models. H-computation therefore restricts the number of possible mathematical universes. However we believe it is possible to somehow ensure within it the extra degrees of freedom that stand out through the Gödel phenomenon.

An argument sketched by K. Svozil [30] also suggests that there is a 'super-halting problem' within H-computation, a fact which may lead to some 'super-Gödel incompleteness' in the corresponding formal theories where their proof machinery has the strength of H-computability. We intend to discuss those questions in a forthcoming joint paper by the authors and K. Svozil.

H-Computation

Let θ be the halting function. Then:

Axiom System 5.1. *The following functions are taken to be H-computable:*

- *The partial recursive functions.*
- *The function* θ. □

H-computable stands for "Hilbert-computable," a name whose motivation stems from the fact that theoremhood in formalized arithmetic becomes decidable. "H-computation," "H-algorithms," and the like have obvious meanings.

Notice that $\theta(m, n) = 1$ if and only if the universal Diophantine polynomial $p(m, n, \dots)$ has roots over the integers and if and only if the (smooth) Richardson transform $\rho p(m, n, \dots)$ crosses the x-axis [6]. ($\theta(m, n) = 0$, otherwise.) So, we take the H-computability of θ to embody a restricted version of our geometrical principle.

We immediately have:

Proposition 5.2. θ *is H-computable if and only if Hilbert's Tenth Problem is solvable with the help of H-algorithms.* □

For the proof, see [6, 9].

H-Complexity

Now embed Turing machine theory into H-computation and formalize everything within T [11]. Consider T-sentences ξ of the form, $\xi =$ "The Turing machine $M_m(n)$ stops and outputs r." $T \vdash \xi$ if and only if $M_m(n) = k$ in the 'external world.'

Now we can obtain a Turing machine M_T such that for some Gödel numbering n_ξ of the sentences ξ of T:

- $M_T(n_\xi)$ converges if and only if $T \vdash \xi$;
- $M_T(n_\xi)$ diverges if and only if ξ is unprovable (either $\neg\xi$ is a theorem of T or ξ is undecidable).

First, construct the corresponding function θ_T. Then $\theta_T(n_\xi) = 1$ if and only if ξ is provable in T, i.e., if and only if we prove the sentence "The Turing machine $M_m(n)$ stops and outputs r," if and only if $M_m(n) = k$. ($\theta_T(n_\xi) = 0$, otherwise.) So we can decide a computation (in this sense) with a single application of the θ function for T. This suggests that *all* Turing computations can be said to be of H-complexity 1 (meaning a single use of θ)—that is, H-complexity equals complete degrees in the arithmetic hierarchy.

VI. Fifth Variation: Traub's Program

The "limits of science" theme lurks behind the foregoing results, but as we already mentioned we still don't know its exact contours—we don't even know if there are clear contours to be perceived. In 1991, Traub sketched [32] a program to inquiry into the limits of scientific knowledge. Here we suggest a few very preliminary steps to advance it.

Let's start with some precise questions [1]:

If one mathematical model of something is computationally intractable, can we prove that all models for it are intractable?

Given a scientific question, construct all the mathematical models for it. Can we prove them to be computationally intractable?

Let us suggest some steps to deal with these queries. Consider (as a guiding example) a physical theory, say, quantum mechanics, as practiced within the domain of everyday, intuitive mathematics. If we restrict ourselves to the first-quantized level, there are no mathematical problems with its formulation. Therefore we can reproduce our first-quantized version of quantum mechanics within, say, ZFC.

From this point, there seems to be two possible roads to follow:

- *The conservative outlook.* Let's suppose that quantum mechanics consists of some "inner core," some irreducible minimal conceptual structure that is somehow reflected in its mathematics. If we take this inner core to be the set of theorems that can be proved within ZFC about the formal counterparts of quantum mechanical objects, then any other theory for quantum mechanics turns out to be an extension of ZFC, which is *per se* undecidable.

- *The alternative outlook.* Now assume we do not need the full power of ZFC in our formal version of quantum mechanics. Suppose that we wish to keep some theorems and discard others. For instance, we wish to keep the full Hahn-Banach theorem, in order to deal with arbitrary continuous spectra, but do not wish to have the Banach-Tarski 'paradox' in our theory. Then if the set of theorems of our new theory is recursively enumerable but not recursive, some kind of undecidability will necessarily crop up. (For instance, there is no decision procedure for theoremhood in the theory.)

Notice [28] that lots of interesting physics can still be done within a classical predicate logic without quantifiers, which is an example for a decidable language. Can we develop a reasonable inner core of quantum mechanics within that language? That's an open question.

If we look towards a still more general kind of investigation we can proceed as follows:

- Start with J. Y. Béziau's "universal logic" [3]. (A universal logic is formal structure that encompasses all formal languages; here a 'deductive system' is some structure where we can define an implicational, deduction-like relation.)
- Try to formulate minimum requirements for a language to include axiomatized theories. (Is that possible?)
- Look for the Gödel phenomenon in those admissible languages.

We can't see much further. Beyond that everything lies open to exploration, like an unconquered frontier [18].

In a certain sense, those things are best left incomplete. We could perhaps play somewhat more with this remarkable word, *incomplete.* But we think it is time to rest our case.

Acknowledgments

Support from CNPq, FAPESP (Brazil) and from FRN (Sweden) is acknowledged. FAD thanks the organizers of the 1995 Abisko meeting on "Limits of Scientific Knowledge" for their hospitality.

References

[1] *Problems on the limits of scientific knowledge*, Abisko (1995).

[2] Bell, J. L. *Boolean–valued models and independence proofs in set theory*, Oxford, Clarendon Press (1985).

[3] Béziau, J. Y. "Universal logic," preprint (1994).

[4] Casti, J. *Reality Rules, II,* John Wiley, New York (1992), pp. 350 ff.

[5] Cohen, P. J. *Set Theory and the Continuum Hypothesis,* Benjamin (1966).

[6] da Costa, N. C. A. and F. A. Doria, *Int. J. Theor. Phys.* **30**, 1041 (1991).

[7] da Costa, N. C. A. and F. A. Doria, *Found. Phys. Letters* **4**, 363 (1991).

[8] da Costa, N. C. A. and F. A. Doria, "Suppes predicates and the construction of unsolvable problems in the axiomatized sciences," P. Humphreys, ed., *Patrick Suppes, Scientific Philosopher*, II, Kluwer (1994).

[9] da Costa, N. C. A. and F. A. Doria, "Gödel incompleteness, explicit expressions for complete arithmetic degrees and applications," *Complexity*, 1995, to appear.

[10] da Costa, N. C. A. and F. A. Doria, "From the halting function to set–theoretic forcing," preprint CETMC–32 (1995).

[11] da Costa, N. C. A., F. A. Doria and D. Krause, "On the satisfiability problem, II" preprint (1996).

[12] Davis, P. J. *Pacific J. Math.* **15**, 813 (1965).

[13] Doria, F. A. in J. Casti and J. Traub, *On Limits*, preprint, Santa Fe Institute (1994).

[14] Feferman, S. et al., eds., *Kurt Gödel: Collected Works*, II, Oxford (1990).

[15] Gibbs, W. "Software's chronic crisis," *Scientific American* **271**, 72 (September 1994).

[16] Heidegger, M. *Was Heisst Denken?*, Max Niemeyer (1971).

[17] Hilbert, D. and P. Bernays, *Grundlagen der Mathematik*, I, Berlin (1934).

[18] Hofstadter, R. and S. M. Lipset, *Turner and the Sociology of the Frontier*, Basic Books, New York (1968).

[19] Horgan, J. "Anti-omniscience: an eclectic gang of thinkers pushes at knowledge's limits," *Scientific American* **271**, 12 (1994).

[20] Leibowitz, R. *L'Évolution de la Musique de Bach à Schoenberg*, Buchet–Chastel, Paris (1957).

[21] Machtey, M. and P. Young, *An Introduction to the General Theory of Algorithms*, North–Holland (1979).

[22] Newman, M. H. A. *Elements of the Topology of Plane Sets of Points*, Cambridge (1964).

[23] Oxtoby, J. C. *Measure and Category*, Springer, New York (1980).

[24] Rice, H. G. *Trans. AMS* **74**, 358 (1953).

[25] Rogers Jr., H. *Theory of Recursive Functions and Effective Computability*, McGraw-Hill, New York (1967).

[26] Siegelmann, H. T. *Science* **268**, 545 (1995).

[27] Soare, R. I. *Recursively Enumerable Sets and Degrees*, Springer (1987).

[28] Suppes, P., personal communication.

[29] Stewart, I. *The Problems of Mathematics*, 2nd. ed., Oxford (1992), pp. 308 ff.

[30] Svozil, K., personal communication, May 1995.

[31] Styx, G. "Aging airways," *Scientific American* **270**, 70 (May 1994).

[32] Traub, J. "What is scientifically knowable," in *CMU Computer Science, a 25th Anniversary Volume*, Addison-Wesley (1991), pp. 489ff.

[33] Yates, F. A. *The Rosicrucian Enlightment*, Paladin (1975).

Chapter 4

THE BARRIER OF OBJECTS: FROM DYNAMICAL SYSTEMS TO BOUNDED ORGANIZATIONS

WALTER FONTANA AND LEO W. BUSS

Overview

Self-maintaining natural systems include the global climate system, all living organisms, many cognitive processes, and a diversity of human social institutions. The capacity to construct artificial systems that are self-maintaining would be highly desirable. Yet, curiously, there exists no readily identifiable scientific tradition that seeks to understand what classes of such systems are possible or to discover conditions necessary to achieve them. Given the ubiquity of such systems naturally and the desirability of self-maintenance as a feature of design, any credible approach to establishing such a tradition merits serious attention.

We have recently developed and implemented a framework for approaching the problem [26, 27]. It is based on the premise that *the constituent entities of a self-maintaining system characteristically engage in interactions whose direct outcome is the* **construction** *of other entities in the same class. Self-maintenance, then, is the consequence of a constructive feed-back loop: it occurs when the construction processes induced by the entities of a system permit the continuous regeneration of these same entities* [87]. We define an *organization* to be the specific functional relationships between entities that collectively ensure their continuous regeneration. A theory of organization, so defined, is a theory of self-maintaining systems. A prototypical instance of entities are molecules. And organisms are a particularly interesting class of self-maintaining systems generated by their constructive interactions. The atmosphere is another example. And so, perhaps, is the sun at the nuclear level.

The overarching long-term goal of our program is to develop a formal understanding of self-maintaining organizations. Our efforts in doing so, which we summarize here, have led us to appreciate a fundamental problem in methodology: the traditional theory of "dynamical systems" is not equipped for dealing with constructive processes. Indeed, the very notion of "construction" requires a description that involves the structure of objects. Yet, it was precisely the elimination of objects from the formalism that make dynamical systems approaches so tremendously successful. We seek to solve this impasse by connecting dynamical systems with fundamental research in computer science, whose theoretical foundations are about "objects" and their constructive interrelations. Our long-term goal, then, becomes equivalent to the task of expanding dynamical systems theory to include object construction, to become what we have come to call *constructive dynamical systems* [26].

I. The barrier of objects

The vast bulk of knowledge base of classical physics has been earned by application of the tools of dynamical systems theory. It began with Newton, and became a powerful tool-kit with Hamilton, Jacobi, and Poincaré. Like all major perspectives in science, its power derives from a useful decision about what constitutes "the system" and what belongs to "the rest of the world." The characteristic feature of dynamical systems theory is to conceptualize "the system" as existing exclusively in terms of *quantifiable* properties (e.g., position, concentration) of interacting entities (real or abstract). The distinction in representing interaction between entities via their properties as opposed to some appropriate theory of the entities themselves will play a major role in what follows. The point is subtle.

In a dynamical system, it is *not* the interacting entities that participate *as objects* in the formal constitution of "the system," but rather their quantitative properties and couplings. As a consequence, interaction is understood as the temporal or spatial change in the numerical value of quantitative variables. This change is captured by a set of (deterministic or stochastic) differential (or difference) equations. The solutions of these equations may then be viewed as a flow in phase space. Analytical and numerical tools exist which permit the characterization of that flow and its change as parameters are varied (e.g., invariant subspaces, attractors and repellors, basins of attraction, bifurcations). In the centuries since Newton, our own century most prominently, the power and efficacy of this cognitive style has been established beyond all question.

The success of this framing in physical systems has fueled an inexorable export of the dynamical systems approach from physics to virtually every domain of biological, cognitive and social science. The record of achievement in these other domains has been mixed at best. To what may we attribute this apparent "limit to scientific knowledge"? A variety of attributions to both specific and general failure are so commonplace that it's tiresome to repeat them once again. Many failures in domains of biology (e.g., development), cognition (e.g., organization of experience), society (e.g., institutions), and economics (e.g., markets) are nearly universally attributed to some combination of high dimensionality and nonlinearity. Either alone won't necessarily kill you; but just a little of both is more than enough. This, then, is vaguely referred to as "complexity."

Laying the blame for scientific limits in this common waste bin, however, has an uncomfortably facile texture. After all, there are examples of wildly successful application of dynamical systems approaches to problems that must have seemed no less daunting at the time than, say, predicting the evolution of the telecommunications market or the global climate looks today. Consider the efforts of R. A. Fisher, whose application of dynamical systems to the problem of combining Mendel's genetics with Darwin's evolution yielded his "genetical theory of natural selection" (i.e., what we now know as the field of population genetics). Fisher accomplished his task via an act of abstraction. His genius was to claim that organisms were an utter distraction and irrelevancy, simply not a part of the problem. The concordance of Darwinism and Mendelism required only a population and genes; the concentration of the latter in the former is the relevant variable, and its behavior may be had by solving differential equations wherein the frequency of a gene is jointly determined by Mendel's transmission rules and Darwin's selection. Fisher's accomplishments belie the conventional mantra of "too complex." Perhaps the "limits to scientific knowledge" are simply a deficit of genius. None of us retain into adulthood a capacity to *seriously* attend for prolonged intervals to an imagined system of abstract entities; whereas every youngster quite seriously attends to the closetful of monsters that appear each nightfall. We rightly celebrate as genius the (first) man who saw genes disembodied from the organism!

In seeking to understand why dynamical systems have had only modest impact in some sciences, the usual explanations are, in some considerable degree, internal to a dynamical systems representation itself. The failures are cast as failures in applying the dynamical systems approach—either a failure of insight in imagining abstract entities appropriate to the system or a failure in tools for the qualitative analysis of high-dimensional, nonlinear differential equations. However, no less real a limit is our ability to stand outside a dynamical systems perspective for a moment and to seriously ask what it is good at and what it is not. Perhaps, then, we might augment the cognitive style itself to render it more tractable in those domains where its achievements have been heretofore limited. This is our intent.

What is left out of thinking about the physical universe as one massive dynamical system with our understanding of it limited solely by insights in framing abstractions well-suited to carving off soluble subsystems? Perhaps what is being too easily overlooked is the fact that dynamical systems never deal with objects themselves [83]. Objects are never represented as entities with a distinct internal structure giving rise to behavior. Rather, objects disappear into arrays of structureless variables confined to holding numerical values that quantify properties of an object class, such as the frequency of a gene, the concentration of a chemical, the density of an electromagnetic field, the position and velocity of an aircraft, the pressure of a gas, the earnings of a firm. The moon, for example, is never represented as an object in the equations that express its orbit; the "moon" is defined as a time-dependent vector of numbers specifying position and momentum. Numerical values are indeed an appropriate abstraction, but *only* as long as objects don't change. Planets interacting gravitationally or Fisher's genes interacting in accord with transmission and selection serve as examples. The situation is quite different when objects possess an internal structure that is subject to change, particularly when that change is endogenous to the universe of objects considered, i.e., when the internal structure of an object causes specific actions to occur that modify (or create) other objects.

Conventional dynamical systems, then, are well-suited for treating changes in the magnitudes of quantitative properties of fixed object species, but ill-suited to address interactions that change the objects themselves. The latter is challenging in the dynamical systems context. The relevant "variables" would have to hold objects, rather than the familiar numerical values. But if the objects become the variables of the system, we would need a "calculus of objects" like we have a differential calculus for numerical values. This places a high premium on the difficult task of abstracting objects without losing the link between their action and structure (i.e., without losing the objects). Perhaps herein lies our seeming deficit of genius.

In Nature, interaction involves objects directly and never by a numerical value describing them. Stepping outside of conventional dynamical systems requires taking this observation seriously. Stated less rhetorically, the occurrence of objects that possess a distinct internal structure of a combinatorial kind has two implications. First, there are substantially more possible objects than can be realized at any given time. It is this which gives meaning to the notion of a *"space of objects."*[1] Second, and most importantly, when the interaction among objects *causes* the construction of further objects, relations of production tied to their internal structure become possible. This never appears in a conventional setting: it can only arise as a consequence of a causal linkage between the internal structure of an object and the actions through which it participates in the construction of others. A theory of such linkage is what a "calculus of objects" would have to accomplish. If we throw out the *constructive* component, we throw out the capacity of a system to *endogenously* induce a *motion* in its "space of possible objects."[2]

What is gained may be seen by analogy to conventional dynamical systems. We imagine construction relations (the analogue of the differential operator) to induce a flow in a "space of possible objects" (the analog of phase space). The intuition is that this flow will have a structure where collectives of objects implementing particular production relations form "attractors" (i.e., "fixed-points," "limit-cycles," and the like) with corresponding "basins." If so, then objects which change one another upon interaction—as surely is the habit of elementary particles, molecules, neurons, firms and governments—have the potential of being characterized and studied as organized collectives of construction relations. The question becomes: Do such organized collectives exist? If so, what are they and what are their properties? Are some self-maintaining, self-repairing, and capable of extension? Is their extension constrained by their internal structure, their history of extension, or both? Are they helpful in filling the void that steadfastly remains in the biological and

---

[1]Eigen [21, 22] has introduced this notion for the special case of nucleic acid sequences—the "sequence-space." Maynard Smith [60] thought of the same in the context of proteins.

[2]Throwing out construction still leaves room for chance events, such as mutation, to induce a motion in object space. The deeper theoretical and conceptual issues arise when the construction of objects derives from the *interaction* among existing ones. not from their variation by chance. The former makes the motion in object space endogenous, while the latter makes it exogenous to the system. Mutation is to construction like perturbation is to dynamics.

social sciences, despite the wholesale importation of dynamical systems approaches into domains so manifestly rich in object construction and transformation?

The issue posed above mandates that the constructive aspect of interaction be brought into the picture. This necessarily requires the representation of objects. In seeking formalisms appropriate to facing the issue of object construction and transformation, one is invariably drawn to the foundations of computation. The computational sciences deal explicitly with syntactical entities, and, thus, with the possible representations of objects and their construction. This defines, then, our specific approach to the general problem. We are obligated to define objects, using formalisms borrowed (at least at the outset) from theoretical computer science, to animate their interaction in an appropriate dynamical setting, and to thereby generate a "motion in a space of objects," the features of which we desire to explore.

All that follows is but a progressive refinement of research tactics we are employing in an attempt to explore this larger question in a specific instance. We concern ourselves with the biological domain, specifically thinking of organisms as self-maintaining chemical collectives. Hence we treat molecules-as-objects and search a corresponding "space of objects" for self-maintaining collectives. We first motivate this choice and show how the simplest abstraction of molecules as agents of construction does indeed generate collectives with a distinctively biological flavor. From this basis, we outline progressive refinements in our abstract chemistry in the form of alternative syntactical systems with the aim of closing the distance from our simplest abstraction of chemistry to something more respectful of chemistry as we know it. After documenting a concrete implementation of the broader perspective in the specific instance of chemistry, we return in conclusion to the larger issues. The reader is urged not to lose sight of the larger goal while immersed in the specific instance: the long chemical excursion is but a logbook of data in support of the utility of the broader view. The proffered "motion in a space of objects" and associated universe of organizations composed of such objects is hardly exclusive to the objects of chemistry and their resultant biological organizations. To make substantive progress—whether in biology or in much of what is beyond biology—we must distinguish and capture the fundamentally different consequences that arise when change is about the objects themselves, as opposed to the magnitude of prespecified quantitative properties describing them.

II. Towards a specification language for chemistry

Our overall goal is more readily grasped and the methodological challenges more concretely framed when stated in the context of a specific class of objects-that-change-objects and a specific organized collective of such objects. Our starting point will be chemistry; the relevant entities are molecules and organizations are self-maintaining chemical collectives. Our motivation in this choice is twofold. First, we chose chemistry because it is solid ground: we know molecules and their interactions far better than any other object class claimed to participate in the construction of self-maintaining organizations (contrast the challenge of molecules versus the challenge of cognitive entities generating markets or firms, for example). Secondly, *we believe that biology, particularly molecular biology, has a pressing need for support from a new kind of theoretical chemistry.* Current quantum and structural chemistry are burdened with information that is not relevant to the molecular biologist. The level of detail and the kind of description offered by these approaches necessarily put the focus on single molecules or individual reactions and away from their functional context within organized systems of molecules or reactions. What chemistry lacks is a high level *specification language* focused on the abstract operational aspect of molecules and capable of describing reaction networks and their *algebraic* behavior. The molecular biologist needs a tool for abstracting molecular actions, for plugging them together (like electronic components), and for generating and analyzing the network closures of these actions under a variety of boundary conditions.

While absence of such a specification will be all-too-apparent to biologists, an example may prove useful to others. Let us consider the role that a yet-unrealized theory of network construction and maintenance might play in understanding how self-maintaining molecular organizations evolve. The scenario is conventional: a mutation occurs, which results in a new_1 gene sequence coding for a new_2 protein whose interaction with the chemical machinery of the cell, set up by the remaining gene products, triggers a cascade of new_3 chemical reactions resulting in a new_4 extension of a metabolic pathway which enables the utilization of a new_5 resource.

Each time the word "new" denotes a different kind of novelty, because each time different kinds of constraints are in effect:

1. $novelty_1$: A sequence is a combinatorial object with the simplest possible structure: a linear concatenation of symbols. The syntactic category of "sequence" entails a space of possible variations. A chain

of 200 positions over an alphabet of four symbols has $4^{200} (= 10^{120})$ realizations—more than the number of bosons in the universe. At this level the generation of novelty is virtually unconstrained. Any random replacement of any symbol at any position yields a new_1 sequence.

2. $novelty_2$: A protein is more than a sequence of symbols. It is a sequence that folds into a shape as a consequence of interactions between symbols along the chain. Three-dimensional space and the nature of intramolecular forces severely constrain which shapes are possible. At chemically relevant levels of resolution these constraints result in considerably fewer stable shapes than sequences. Not every $novelty_1$ is a $novelty_2$.
3. $novelty_3$: The types of functional groups and their disposition within a molecule define its "domain of interaction"—its capacity to participate in specific chemical action (i.e., the breaking and making of bonds). $Novelty_3$ is a matter of chemistry.
4. $novelty_4$: The constraints and opportunities of interaction within a given network of chemical pathways determine which new_4 network roles a new_3 molecular agent can participate in. How (or, even, whether) a network forms depends on its molecular components, the types of reactions induced by them, the connectivity of these reactions and their kinetics.
5. $novelty_5$: The $innovated_4$ metabolic network is characterized by constituent molecules and their relationships. What is regarded, however, as a new_5 "resource" or as new_5 "waste" is a matter of the coupling between this network and other such networks either within the same, or between it and other, levels of biological organization. Indeed, it is the joint construction and maintenance of a chemical reality composed of a large number of linked metabolic networks which defines the biotic element of an environment.

It is plain that $novelty_5$ cannot occur unless $novelty_1$ occurs. There is, however, a gap between $novelty_1$ and $novelty_5$ that theory is presently unable to bridge. We perfectly understand the "abstract space of possibilities" for $novelty_1$: it's the space of words over an alphabet. Yet we have basically no clue as to even the nature of the abstract space of possibilities for $novelty_5$. The two loose ends of the problem circulate in biology under the key-words "genotype" ($novelty_1$) and "phenotype" ($novelty_{i>1}$). The evolutionary process is perceived roughly as the conjunction of two factors: the modification of "phenotypes" by chance events at the level of "genotypes," and a dynamics which results in the selective amplification of "genotypes" based on the differential reproductive success conveyed by their "phenotypes." $Novelty_1$ is as simple as a throw of the dice. However, once it has occurred, we lack utterly the capacity to assess its likelihood of giving rise to $novelty_5$. Yet, this likelihood is defined by a molecular society, the constituent interactions of which have a lawful—even largely deterministic—character grounded in physics and chemistry. Might there not be an abstraction of chemistry appropriate to such questions?

We claim that *any serious attempt to mathematize such questions requires an abstract characterization of chemical processes.* This stance defines our more specific goals:

- to develop an "abstract chemistry" in which molecules are viewed as computational processes supplemented with a minimal reaction kinetics, and

- to develop a theory of the self-organization, maintenance, and variation of networks based on such processes.

Situating these specific goals in the broader perspective, we believe that an adequate abstraction of chemistry is as crucial in extending the theoretical foundations of biology as was an adequate abstraction of motion in founding a formal basis for physics. The parallel, however daunting, is one we make seriously. Roughly, "motion" in physics is conceived as the temporal change in the value of a state variable (the position, say). This motion is formalized by infinitesimal calculus; a theory of the derivative d/dt. We would like to think of chemical reactions as a kind of "motion" as well—but as a motion in a *space of objects*. The key difference is that, mathematically, such objects are not numerical quantities, they are syntactical entities, to wit: molecules. The chemist denotes that motion with "$\longrightarrow$", as in $CH_3OH+CH_3COOH \longrightarrow CH_3COOCH_3+H_2O$. The objects on the left are replaced by those on the right. But these may interact further with other molecular agents present in the reaction vessel or the cell, thereby keeping its contents changing over time, that is, "moving in object space." What is needed is a theory of the motion generator "$\longrightarrow$" competent to define a universe of self-maintaining organizations of such objects. If the broader perspective is correct and the specific implementation sufficiently exact, within this universe of organizations will be found specific organizations known to us as living biological systems. The reader will find grounds in our simplest implementation in support of the validity of the broad claim, but will find manifest inadequacies in the precision of the specific chemical implementation presented in section II.1. Optimism inspired by success in the former has motivated the series of refinements summarized in sections II.2 and II.3 in attempt to improve upon the deficits of the latter.

II.1. Minimal Chemistry Zero

II.1.1. Ontological commitment, resultant metaphor and formal representation

In view of the above discussion, the principal question is "how to frame chemistry?" The problem is one of focus—how close-up? how distant?—and one of scope—how wide? how narrow?. We are forced to make ontological choices. To begin with, we choose an absolutely minimalist view of chemistry.

1 - Syntactical. Molecules are treated as discrete structures of symbols, defined inductively. A molecule is an atom or a combination of atoms or molecules.

2 - Constructive. Reactions are seen as events where such symbolic structures "interact" to construct new symbolic structures.

3 - Substitution. The basic mechanism of a reaction is the exchange of one group of symbols (a substructure) by another, i.e., a substitution.

4 - Equivalence. Different combinations of reactants can yield the same product.

5 - Deterministic. When particular functional groups in a molecule initiate a reaction, the product is determined.

This minimalist view coincides with the description of a mathematical function as a rule, rather than a set. In the former case a function is a suite of operations that generate an output when applied to an input. In contrast, the latter case views a function as a look-up table, i.e., a set of input/output pairs. To express rules, a syntax is needed (point 1). Functions-as-rules can be applied to arguments which can themselves be functions, returning a new function as a result (point 2). For example, take a polynomial and "apply" ($\circ$) it to another one: $(x^2+4x+2) \circ (y-1) \longrightarrow (y-1)^2+4(y-1)+2 \longrightarrow y^2-2y+1+4y-4+2 \longrightarrow \cdots \longrightarrow y^2+2y-1$. This also illustrates that the process of evaluating the application of a function to an argument is done by repeated substitutions, where the formal variable of a function is replaced by the literal text of the argument (point 3). The schemes which govern this process define a calculus. Furthermore, different function/argument combinations can return the same result. Trivially, $7+3 = 20/2$, but "$7+3$" is not the same object as "20/2". To justify the equality a syntactical manipulation—a computation—must occur that puts each object into its canonical form: $7 + 3 \longrightarrow 10 \longleftarrow 20/2$ (point 4). Furthermore, the application of a particular function to a particular argument always yields the same result (point 5), although it may proceed via different routes (depending on which subexpressions are evaluated first).

This, then, is summarized by the following metaphor:

chemistry	......	**a calculus**
physical molecule	......	symbolic representation of an operator
molecule's behavior	......	operator's action
chemical reaction	......	evaluation of functional application

To put the metaphor to work, it must be made precise. Any formal system that is a candidate for an abstraction of chemistry at this level must make the same ontological choices. The *only* canonical system known that formalizes the notion of a function as a rule, and is both based on substitution and naturally yields a theory of equality is the λ-calculus, which was invented in the 1930s [12–14, 82]. It has since become of foundational importance in the computational sciences. We find it remarkable that there is a system at all—and even such a central one—that fits so well. The correspondence with λ-calculus will later enable substantial refinements of the chemistry/calculus metaphor, thereby providing an *ex post* justification of this choice.

Although not strictly necessary to grasp most of the remainder of this chapter at an intuitive level, the reader unfamiliar with λ-calculus is invited to consult[3] [89], Appendix A, for more details.

---

[3]To keep this chapter within a reasonable size, we omit appendices A, B and C whose purpose is to serve as a guide to λ-calculus, type theory, and proof-theory, respectively. These materials are available in conjunction with the main text as IIASA Working Paper WP–96–27 or as Working Paper #41 from the Yale Center for Computational Ecology. They are also obtainable from the World Wide Web at `http://peaplant.biology.yale.edu:8001/alchemy.html` or `http://www.iiasa.ac.at/docs/IIASA_Publications.html`.

II.1.2. Model

Motivated by biological problems akin to that sketched at the beginning of section II, we have developed and implemented a toy model aimed at exploring the conjunction of the two interaction modes—construction and dynamics—introduced in section I.

Our abstract molecules are symbolic operators expressed in the λ-calculus. We consider a "flow reactor" of N such abstract molecules, each one a λ-expression. In this setting a given expression may occur in multiple instances, just as in a test tube a number of molecules may be instances of the same chemical formula (see Figure 1). We think now of the expressions as if they were particles floating within a well stirred solution where they collide at random. Upon contact, two expressions interact by functional application, such that one expression assumes the role of operator, and is applied to the other expression which assumes the role of argument. The evaluation of this interaction yields as its result a new expression. Thus, the canonical calculus realizes the desired constructive component—collisions (i.e., "applications" followed by reduction to normal form; see [89], Appendix A.2) *are* production relations among abstract molecules—and these occur in a particular dynamical setting (i.e., the flow reactor) such that construction is coupled to changes in the concentration of the expressions.

We omit details of our implementation not essential for the purpose of this overview. They can be found in [26, 27]. One issue is, however, immediately germane. Recall that we wish to induce a "motion in object space," with that motion settling upon self-maintaining systems of objects. To achieve the latter, we impose a generic selection constraint on object motion through a choice of reaction kinetics and a restriction on reactor size. Specifically, we make two assumptions:

- Reactants are *not* used up in a reaction:

$$A + B \longrightarrow C + A + B, \tag{1}$$

where C is the normal form result that is contingent on the application of operator A to argument B: $(A)B \rightarrow C$ in λ-calculus. In this way the total number of expressions increases by one with each reactive collision.

- Each time a new expression has been produced, a randomly chosen one, X, is removed from the reactor:

$$X \longrightarrow \emptyset. \tag{2}$$

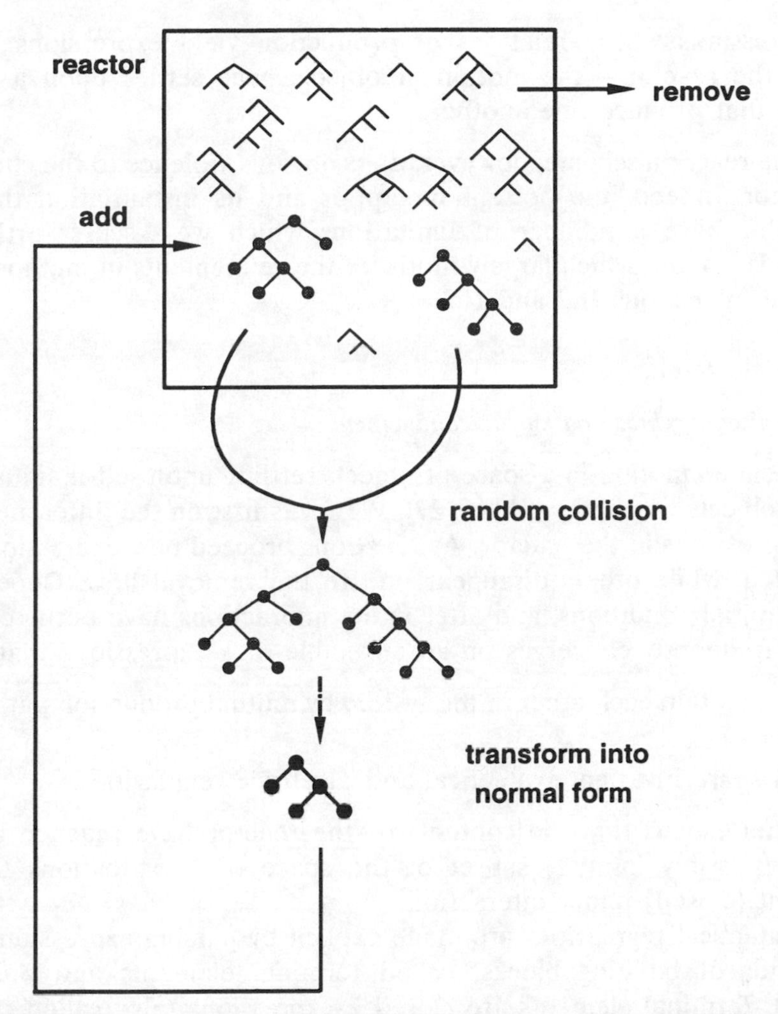

Figure 1. A λ-calculus flow reactor. Two expressions A and B are chosen at random and a new object, $(A)B$, is constructed by "application" (see [89], Appendix A.2]). Putting $(A)B$ into its normal form by β-reduction (see [89], Appendix A.2), effectively decides which object species (i.e., "stable molecular formula") the new object is an instance of.

The overall number of expressions N is thereby kept exactly constant. This means that each expression has a finite life time, even though it is not consumed at the moment of a reaction. Moreover, since any two expressions interact to produce a particular third expression with a frequency proportional to their concentration, the reaction scheme together with fixed reactor size act to favor convergence to a population of expressions whose relations of production yield expressions extant within the reactor – the motion in object space settles upon a set of objects that produce one another.

The reaction scheme, however, does obvious violence to the chemical metaphor. Indeed, the present metaphor and its instantiation through λ-calculus have a number of limitations which we discuss further in section II.1.4 and which largely motivate the refinements in methodology outlined in sections II.2 and II.3.

II.1.3. Main results

Self-organized algebras and kinetic confinement

The intended motion-in-a-space-of-objects settling upon self-maintaining-sets-of-objects was observed [26, 27]. We focus first on the different kinds of λ-expressions in the reactor. As reactions proceed new expressions are generated, while others disappear due to the removal flow. Depending on the initial conditions, and after many interactions have occurred, the system frequently converges on an ensemble of λ-expressions that

(i) maintain each other in the system by mutual production pathways, and that

(ii) share invariant syntactical and algebraic regularities.

The latter means that the contents of the reactor have reached a particular (possibly infinite) subset of the space of λ-expressions that is invariant (closed) under interaction.

Syntactical regularities are made explicit by parsing expressions into two kinds of building blocks, called terminal elements and prefixes[4] [26, 31]. Terminal elements are closed λ-expressions (also called combinators, see [89], Appendix A). Prefixes are not complete λ-expressions. However, prefixes form closed expressions when they precede a terminal element. The invariant subspace contains only expressions that are made from a characteristic set of such building blocks.

Algebraic laws are a description of the specific action(s) associated with each building block. This action may depend on the context of a building block within an expression. The characterization of the functional relationships among the blocks yields a system of

[4] We define a terminal element to be the smallest closed subexpression reading from the end of a λ-expression. A prefix is a smallest closed sub*structure.* It need not be (and typically is not) a well-formed expression.

rewrite equations [50]. This system can, in many cases, be exhaustively specified using Knuth-Bendix (and related) completion techniques [51, 52][5]. Rewrite systems which complete, permit a finite specification of all interactions among the expressions of the subspace. They implicitly determine a grammar for its (normal form) expressions.

The rewrite system cast in terms of building blocks is a *description of the converged reactor system in which all reference to the underlying λ-calculus has been removed.* In other words, the generic λ-calculus can be replaced by another formalism specific to the self-maintaining ensemble of expressions in the reactor, that is, a particular algebraic structure.

The expressions of the invariant subspace are the carrier set of the algebra. Very often, but not always, that set is infinite. Although the reactor has only a very small capacity (1000 or 2000 expressions), the algebra persists through a fluctuating, yet stably sustained, finite set of expressions. This occurs whenever the connectivity of the transformation network is such that it channels most of the production flow to a core set of expressions. This we call *kinetic confinement*. An example is shown in Figure 2.

Organization

The main conceptual result is a useful working definition of what we mean by an "organization": *an organization is a kinetically self-maintaining algebraic structure.* Self-maintenance has here two aspects which reflect the two modes of interaction: (i) algebraic, a network of mutual production pathways that is a fixed-point under applicative interaction, and (ii) kinetic, the concentrations of the expressions in the network core are maintained positive. The former is a necessary, but not a sufficient condition for the latter, i.e., a network can be algebraically a fixed-point—every expression being produced within the network—but its particular connectivity may not suffice to sustain non-zero concentrations of its core components under flow-reactor conditions as specified by Eqs. (1) and (2).

Organizations of differing algebraic structure are obtained by varying the set of λ-expressions used to seed the reactor. An infinity of such

[5]Some rewrite systems induce an infinite recursion and defy completion. In our system, this is manifested as building blocks whose action upon one another is to generate new building blocks with the same property. The failure of some rewrite systems to complete is a consequence of the unsolvability of the universal word problem.

organizations are possible. Developing a taxonomy of their structure and properties remains a long-term goal of our program.

Self-repair and constrained variation

Two prominent properties of these organizations are their resilience to the subtraction of existing components and resistance to the addition of new expressions. Organizations often repair themselves following removal of even large portions of their component expressions. Some organizations are even indestructible: they regenerate themselves from any component.

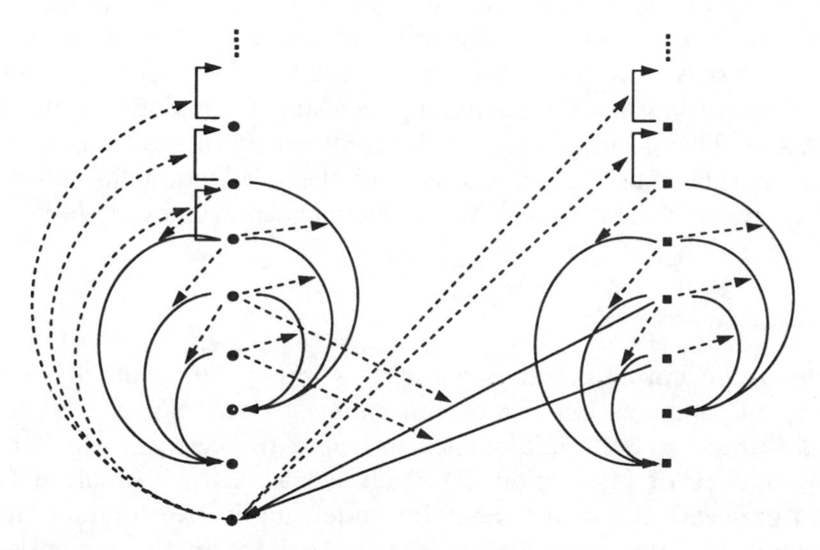

Figure 2. A simple self-maintaining organization. The dots (left) and the squares (right) represent λ-expressions with a particular grammatical structure. They are made of one prefix $[\lambda x.(x)]$ and two terminals $[T_1 \stackrel{\text{def}}{=} \lambda x.x$ and $T_2 \stackrel{\text{def}}{=} \lambda x.\lambda y.(y)\lambda z.(z)x]$. From bottom to top, the dots (left) are expressions consisting of an increasing number of prefixes (starting with 0 at the bottom) terminated by T_2. The same holds for the squares (right), except that they are terminated by T_1. A solid arrow indicates the transformation of an argument (tail) to a result (tip) by an operator (dotted arrow). For clarity, only a subset of the possible interrelations is shown. Notice the connectivity enables kinetic confinement. Most transformations yield objects at the bottom (leading to an increasing concentration profile from top to bottom). Some operations, however, yield objects up the "ladder," thus establishing self-maintenance. Both syntactical families depend on each other for maintenance as indicated by the "cross-family" connections.

The reason for this robustness is the existence of *generators* of the algebra. These are subsets of expressions whose repeated interactions rebuild piece by piece the entire organization; if they are retained, the system regenerates.

The link with algebra also clarifies an organization's response to the addition of new expressions, but for a different reason. The grammatical and algebraic invariances can be viewed as abstract *boundaries* of the organization. They determine membership. An expression that does not conform with that organization's particular grammar cannot be a member of the organization. Despite having an independent description, an organization is embedded in the larger λ-universe, and a non-member expression may perturb the organization algebraically (and grammatically), generating further expressions "outside" of it. The perturbing expression can, in some cases, be stably maintained leading to an *extension* of the original organization. Alternatively, the perturbing expression may be diluted out of the system leaving the organization unaltered. The algebraic relationships which define an organization also determine *specific* opportunities for its extension. Biological interpretations are many. As but one example, Morowitz suggests that nonenzymatic precursor networks of the cellular core metabolism have evolved via distinct extensions [72].

Organization within organization

Organizations can have a quite complex substructure. To explain what we mean by substructure we need two iterated mappings. One is an "expansion" of a set $\mathcal{A}_i$ of expressions:

$$\mathcal{A}_{i+1} = \overline{\Omega}(\mathcal{A}_i) \stackrel{\text{def}}{=} (\mathcal{A}_i \circ \mathcal{A}_i) \cup \mathcal{A}_i \tag{3}$$

where $\mathcal{A} \circ \mathcal{B}$ means the set resulting from applying every expression in $\mathcal{A}$ to every expression in $\mathcal{B}$. The other is a "contraction" of a set:

$$\mathcal{A}_{i+1} = \underline{\Omega}(\mathcal{A}_i) \stackrel{\text{def}}{=} (\mathcal{A}_i \circ \mathcal{A}_i) \cap \mathcal{A}_i \tag{4}$$

Given an organization O generated in our reactor, we take each expression i in O and iterate $\overline{\Omega}$ T times to obtain an expansion of i: $\mathcal{B} = \overline{\Omega}^T(\{i\})$. After this we contract $\mathcal{B}$ until we have found a fixed point: $O_{\{i\}} = \underline{\Omega}^{T+1}(\mathcal{B})$. In summary, $O_{\{i\}} = \underline{\Omega}^{T+1}(\overline{\Omega}^T(\{i\}))$. If $O_{\{i\}}$ is not empty, we have obtained a self-maintaining suborganization contained in O that has been generated by the single expression i.

The relationships between all suborganizations generated by individual expressions of an organization can be visualized in a lattice partially ordered by inclusion. An example is shown in Figure 3. The topmost node represents the entire organization. It is a combination of 11 suborganizations located at the next lower level in the diagram. The left-most suborganization, for example, is an extension of the organization below it (darker node), which in turn is an extension of the black node. Since the black node is contained in a number of organizations above it, these organizations necessarily overlap (i.e., they share some members). The bottom node is a small closed self-maintaining set contained in all others. Despite its apparent complexity, only three interaction laws involving only one terminal element and two prefixes are required to describe the system.

The substructure of an organization reflects only the algebraic aspects of the organization. Any physical realization of such an organization is also a matter of dynamic stability. Structure and dynamics jointly define organization-specific properties with respect to robustness and evolvability.

Higher-order organizations

We can combine *disjoint* organizations that have been obtained independently. In some instances they build a stable higher-order organization that contains the component organizations in addition to a set of products arising from their cross-interactions. This set is not self-maintaining, yet it is crucial in stabilizing their integration into a new unit. We call such a set of objects a *glue.* Biologists will recognize this as an issue of some importance in history-of-life [8, 61], e.g., the mitochondria and chloroplasts of eucaryotic cells are descendants of cells with an independent procaryotic ancestry.

Copy functions and the emergence of organization

Our model universe invites experimentation on the conditions which facilitate or impede the emergence of organization. An example of one such condition involves the role of replicating objects, that is, λ-expressions that copy (see Figure 4).

Replication is a term usually used to denote an autocatalytic *kinetic* role, i.e., an agent whose change in concentration is proportional to its own concentration. In addition to its kinetic aspect, the present model makes the *operational* role of a replicator explicit. A replicator is the fixed-point of some interaction. Then f is a replicator if the system

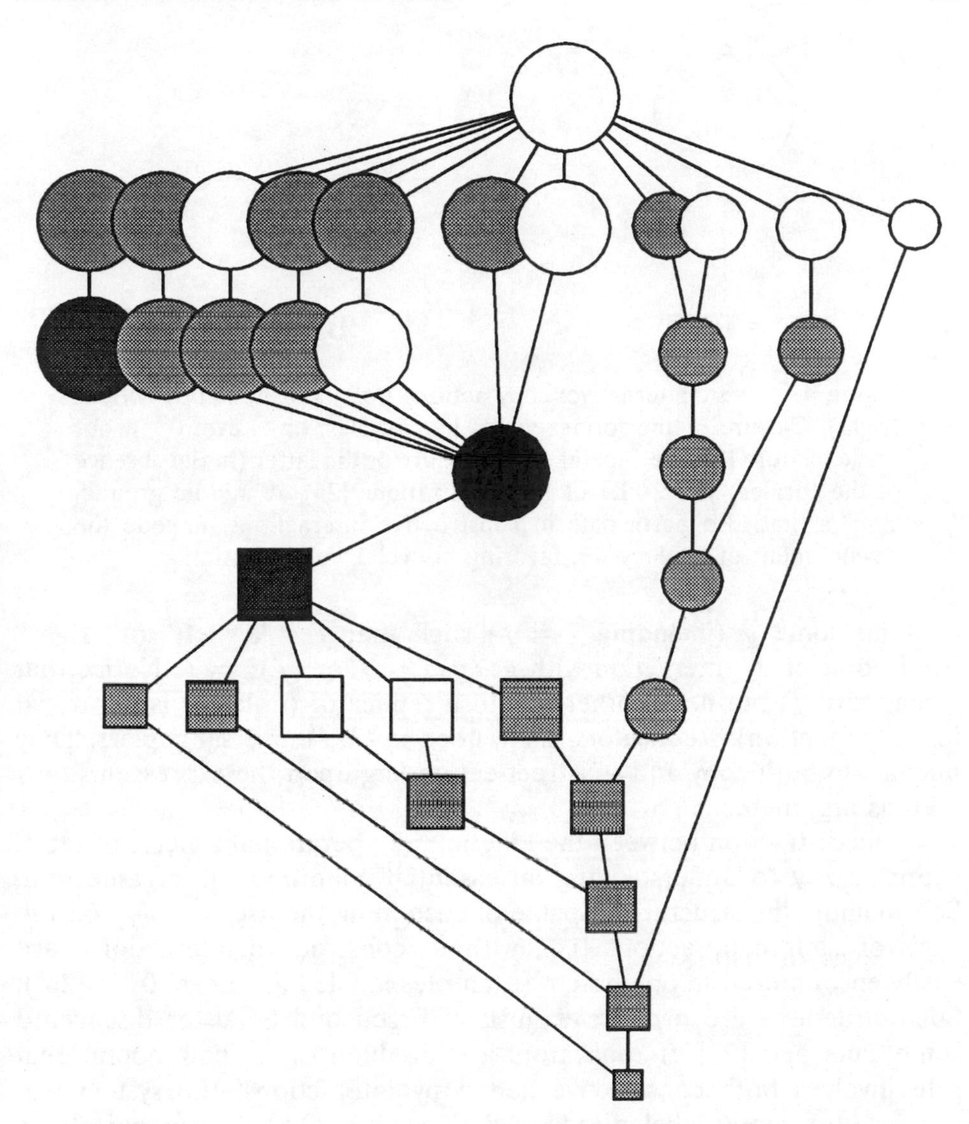

Figure 3. The substructure of an organization. Each node (circle or square) represents a self-maintaining set. Circles denote self-maintaining subspaces with a potentially infinite number of expressions, while squares represent finite self-maintaining subsets. When two nodes are connected by an edge, the lower one represents a set that is contained in the upper one. The size and grey level of a node reflects that node's share of the overall diversity and total number of expressions in the reactor, respectively. See text for further discussion. (Figure and analysis courtesy Harald Freund.)

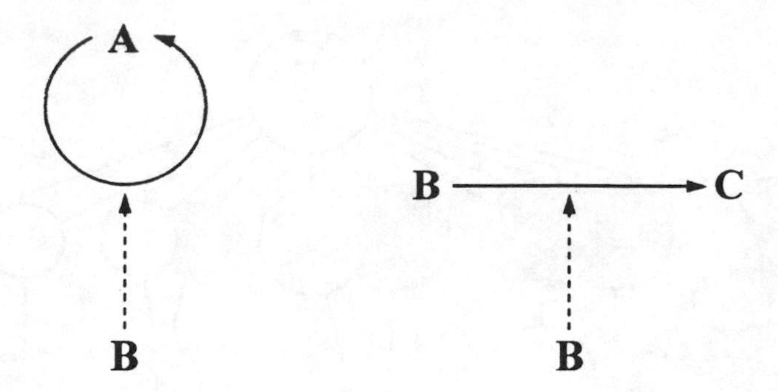

Figure 4. A basic alternative: copy actions (left) and non-copy actions (right). Closure of the former yields hypercycles, or "Level 0" in our nomenclature [26]. Self-maintaining closure of the latter (in the absence of the former) yields "Level 1" organizations [26]. A middle ground, copiers that also participate in constructive interactions, impede the development of hypercycles, favoring "Level 1" organization.

contains some g (including $g = f$) such that f is a "left" or "right" fixed point of its interaction with g: $(g)f = f$ or $(f)g = f$. Notice that g may turn f, but not another h, into a replicator (unless g is the trivial identity function). Replicators, then, need not be universal copiers. They may act to both copy and construct depending upon the expressions they take as arguments.

The distinction between the kinetic and operational aspects of replication is key to understanding an essential condition for organization. Self-maintaining structures capable of sustaining themselves *solely* on the basis of their copy actions (i.e., without constructive interactions) are easily encountered in our system (examples labeled as "Level 0" in [26]). Such structures are *hypercycles,* just as Eigen and Schuster discovered some time ago [23]. If replicators are disabled or if their operational role involves both constructive and copy interactions, the system will organize (examples labeled as "Level 1" in [26]). "Level 1" organizations differ fundamentally from hypercycles in their self-repair and extensibility properties.

II.1.4. Main limits

The results summarized above clearly illustrate that the merger of a dynamical system (the flow reactor) with a universe of objects that entertain constructive interrelations (λ-expressions) does indeed achieve the desired objective. A motion-in-the-object-space is induced, such that self-maintaining structures, characterized by an invariant pattern of transformations, arise. Moreover, these organizations possess properties – regeneration, structure-dependent extension, complex substructure, capacity for hierarchical nesting—akin to properties of living organisms. Yet, λ-expressions are far from molecules and our organizations far from organisms. The major limitations of Minimal Chemistry Zero (MC0) are enumerated below.

1. **Shape:** Molecules interact selectively. This is violated in MC0 because λ-operators can act on one another indiscriminately.
2. **Symmetry:** Reaction is a symmetric event. Violated in MC0, because functional application is not commutative.
3. **Mass action:** With respect to a reaction event, molecules are resources and are used up. Furthermore, atom types and number are conserved during a reaction event. This principle is violated twice in MC0, first by the kinetic scheme (1), and second microscopically—which is far more serious—by the multiple occurrence of the same bound variable in λ-expressions. To make the latter clear: when supplying the argument 5, say, to the function $f(x) = x^2 + 2x + 3$, the 5 gets used twice; once when substituting in x^2 and once in $2x$. Where does the second 5 come from? In chemistry, a reaction has only as many atoms as are present in the reactants.
4. **Reaction classes:** Chemical reactions proceed according to a variety of distinct schemes, such as substitutions, additions, and eliminations. In particular, individual reactions can yield several molecules on the product side. Violated microscopically in MC0, because application in λ-calculus yields at most one normal form (product). (Note that the reaction scheme (1) is an exogenous condition we impose.)
5. **Rate constants:** In chemistry reactions proceed with different velocities, which leads to a separation of time scales in reaction networks. This is violated in MC0 because every reaction event has the same unit rate constant.

These limitations are substantial and motivate the improvements to which we now turn.

II.2. Minimal Chemistry One

We consider an extension of MC0 designed to address the issue of "shape" (item 1 in subsection II.1.4). Pure λ-expressions are strings of characters that represent functions with no specific domain of definition (i.e., they can act on any expression). Shape enforces a specificity upon interaction.

II.2.1. Shape and action

The virtues of MC0 lie in the transparency of λ-calculus and the connections its use provides to abstract algebra and rewrite systems. It is difficult to imagine how a 3-dimensional interpretation could be given to the actions of λ-expressions in a canonical way. An explicit spatial representation would seem to be required. However, a price for capturing shape would surely be paid in transparency of the resulting model. Might there not be an abstraction of shape that evades the costs of explicit spatial imitation?

Molecular shape derives from a self-consistent balance of nuclear and electronic motions influenced by each other's field. At the same time the resultant distribution of electronic and nuclear densities gives rise to specific chemical properties. In this sense shape and chemical action are two sides of the same coin. In a slightly more abstract sense, the specificity of chemical action between molecules results from (i) the *complementarity of chemical properties* between reacting functional groups, and (ii) their spatial disposition. The first aspect means, for example, that an electron donor group on one molecule must meet an electron acceptor group on the other for an action between them to occur. To put it in a cartoonish way, chemical complementarity emphasizes that action occurs when one functional group is of the type ``if I'm given an *x*, then I yield a *y*'', while the other group is of the ``I'm an *x*''-kind. If the latter were an ``I'm a *z*'', no action would take place. It is clear at once that interaction selectivity, though invariably tied to space in real chemistry, does not require space to be expressed abstractly.

When a reaction involves more than one chemically complementary group, their spatial disposition further contributes to specificity by excluding those reaction partners that have the right groups at the wrong places. However, this is a combinatorial aspect that is neither unique to spatial extension, nor one that fundamentally alters the nature of specificity caricatured above.

A rather different issue is raised by non-reactive molecular interactions based on shape. There, the *geometric* aspect of spatial form is essential in giving rise to supramolecular morphologies, such as membranes or viral capsids. This aspect necessarily escapes a formalization cast in a non-geometrical syntactical system; it is as much outside the calculus-metaphor as is the flow-reactor kinetics. At this stage of our program, however, we dispense from further physical embeddings (beyond kinetics); our interest being in transferring as much as possible of what appears to be physical to an abstract computational domain.

Two aspects of molecular form, shape-as-conditional-action and shape-as-geometry, are together responsible for chemical interaction specificity. Here we formalize only the first aspect, taking the stance that it is not the molecule's shape-as-a-coordinate-list that counts, but rather how the spatial configuration is parsed into basic reaction classes. (A virtuoso synthetic chemist looks at a molecular configuration in much the same way that a grand master looks at a chess configuration, perceiving the molecule in terms of what can be *done* with it, i.e., which features can be exploited to make or break bonds with respect to a synthesis goal.) Thus, to the extent that shape is abstracted as a suite of lawful restrictions on permissible actions, it is plausible to capture its role by imposing a suitable discipline upon λ-interaction. This is done by augmenting the notion of function in λ-calculus with the constructive analog of a "domain of definition" and a "range."

II.2.2. What is a type?

Minimal Chemistry One employs the use of typed versions of λ-calculus, where the system of types serves as an abstraction of restrictions on chemical action. Here, we briefly introduce the notion of a "type." A more detailed but still expository overview can be found in [89], Appendix B. For a rigorous treatment the reader should consult the literature [9, 10, 38, 55, 77].

A type is a statement about overall action. To appreciate this, consider an *untyped* universe, such as a computer at the level of memory cells [10]. It appears as an unstructured array of binary strings undergoing transformations. When looking at these strings we typically have no way of telling what is being represented. In contrast, a *typed* universe, such as a programming language, provides frames of interpretation for the digital contents of computers by imposing a kind of semantics defining intended use. Such frames work by offering a repertoire of behavioral *types*, such as variables, arrays, pointers, procedures, and control structures. Furthermore, variables themselves are often distinguished according to the *type* of value they are meant to hold: boolean, real, integer, character, and so on. *The effect of such constructs is basically to enforce a discipline of interaction.* For example, the interpreter of a programming language rejects the application of a function that removes blanks from character strings to an "inappropriate" object such as a vector of numbers. In essence, a type is but an object's "interface" that regulates with what it may communicate.

In programming languages a type serves as a *specification,* that is, it provides partial information about what an operator (a program) abstractly does. In chemistry, however, there are no external reference frames, no intentionally defined "integers," "character strings," or "vector products." The lawful behavior of chemistry is internally defined by the underlying physics. It is, therefore, important to understand that the abstract notion (and theory) of types is *independent* of any particular meanings. A representation of chemistry at a chosen level of resolution could be defined by a repertoire of primitive objects with assigned behaviors. Their internal structure is suppressed, and the behaviors are defined reciprocally. This is what computation theory calls an "abstract data type." Primitive objects of this sort could be atoms, or functional groups, such as hydroxy, amino, carbonyl groups, etc., or they might be further abstracted entities, those which carry "oxidizing" behavior, others with "reducing" behavior, "acid" behavior, "base" behavior and so on. The action of a chemical group as a primitive could be specified by indicating which other groups it interconverts, without indicating how this is done (e.g., mapping a keto group into a hydroxy group under certain conditions). To turn this into a chemistry a mechanism for building complex objects from primitive ones and for interconverting them is needed. That is what typed λ-calculus provides. Admitting primitives with a specific behavioral interpretation amounts to defining constraints as to which objects can be built and, therefore, which reactive combinations are possible.

II.2.3. Improved metaphor

We have implemented a simple standard type system for λ-calculus [15, 63], following the path laid out by a very useful prototype [59, pp. 97–113]. The system is explained in [89], Appendix B. Here we emphasize only those conceptual features that are important for our chemical agenda.

• **Syntactical structure and type are coupled.** A type is not arbitrarily attached to a λ-expression. It is derived from its syntactical structure by means of inference rules in a process called *type synthesis*. If an expression is modified, its type changes accordingly. The requirement to possess a type constrains the syntactical structure, and excludes some of the expressions that were possible in the untyped case. These constraints are interpreted to reflect the fact that a molecule's specific domain of action is based on its structure and composition, and that the properties of atoms constrain what kinds of molecules there can be.

- **Type polymorphism and boundary conditions.** Types can convey different degrees of specificity. A particular type may constrain an expression to act on one sort of argument only, while another may not discriminate at all. This is *type polymorphism* (see [89], Appendix B). The degree of polymorphism is controlled by assigning *basis types* of chosen specificity to the variables (and constants, if any) of the λ-system. *The set of basis types constitutes a new boundary condition*. It permits the tuning of the overall reactivity of our abstract chemistry and the definition of primitives with specifically chosen interrelations. It is in the definition of the basis set that an abstraction of molecular shape-as-conditional-action (or any other intended restriction upon action) succeeds or fails.

- **Interaction specificity.** An expression that represents a map sending objects of type τ into objects of type σ, can act upon arguments of type τ. To decide when an interaction can occur is not as trivial as it seems. The type-expression can be viewed as describing a domain whose size reflects its degree of polymorphism. Whether the polymorphic types of two colliding λ-expressions match properly is not a mere syntactic comparison, but involves detecting whether one type is an instance of the other. The decision procedure is outlined in [89], Appendix B.

The present formalization treats the abstract essence of "shape" as a statement about a molecule's domain of action. It bears emphasis that in this formalization the λ-term continues to be the object corresponding to the physical molecule. The type-expression derived from the λ-term is but a device to enforce an interaction specificity. If the types of two colliding objects permit their interaction, the syntactical manipulations follow the λ-calculus. It is good conceptual hygiene not to confuse the type with the object. This is plain in chemistry, the shape of a molecule is not the molecule.

These features, in aggregate, define the metaphor underlying Minimal Chemistry One, shown in the table on page 82.

II.2.4. Model and preview of results

The reactor with Minimal Chemistry One is schematically shown in Figure 5.

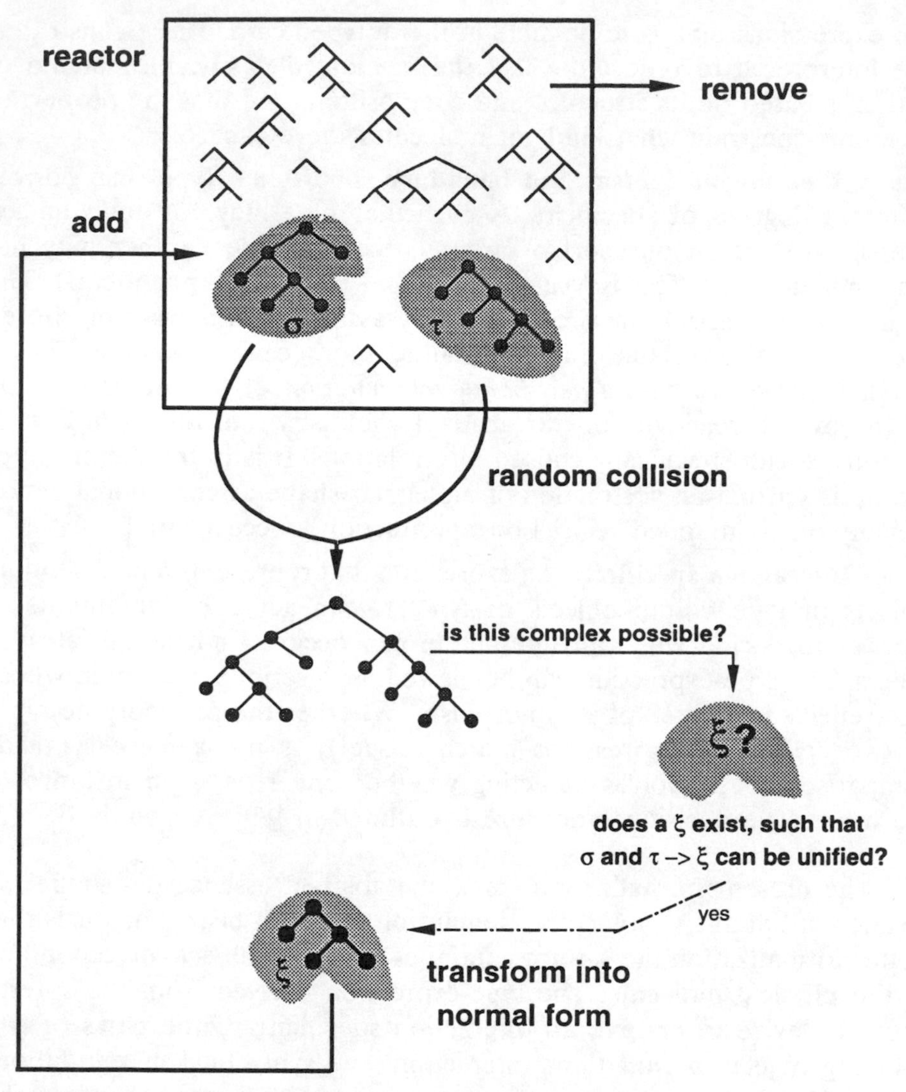

Figure 5. The λ-calculus flow-reactor with function-particles that discriminate among interaction partners on the basis of a type system. Two randomly chosen expressions with types σ and τ (represented as shaded regions) collide. The validity of the interaction complex depends on whether a type can be assigned to it. The procedure is explained in [89], Appendix B. If the interaction complex is typable, the reaction proceeds by normalizing the complex. Otherwise, the types σ and τ are incompatible for interaction, and the collision is regarded as elastic.

chemistry	······	typed calculus
physical molecule	······	symbolic representation of an operator
molecule's behavior	······	operator's action
specificity of interaction	······	type discipline
chemical reaction	······	evaluation of functional application

Our results with the MC1 model have yet to be exhaustively reported in the primary literature and, accordingly, we will not provide as detailed a summary of results as presented for MC0. The major consequence, however, of the improved model is that organizations are once again achieved and display properties akin to those documented for MC0. Organization, however, is considerably more difficult to achieve than in the untyped case, as may be expected by a restriction on interaction. The degree of difficulty is related to the degree of polymorphism, and, therefore, to the type basis.

II.3. Minimal Chemistry Two

Only one of the limitations inherent in Minimal Chemistry Zero, Section II.1.4, is addressed by our abstraction of shape as a lawful discipline upon interaction and our accompanying implementation of that abstraction in typed λ-calculus. Problems with MC0 regarding symmetry, resource accountability, reaction classes and rate constants remain in MC1. Indeed, one might even contend that our notion of types-as-shape is not mature *until* these problems are solved—that is, until we succeed in defining a basis set that generates, for example, the appropriate classes of reaction in some restricted chemical domain.

The issue is one of the level of abstraction we chose. A critic might well contend that our level of abstraction is so high as to willfully preclude eventual maturation from an abstract to an actual chemistry. This, however, would be a misreading of our intent; see, for example, [84]. The retention of a high level of abstraction in the transition from MC0 to MC1 is anything but a resistance of the actual. Rather it represents a strategic claim that the benefits of a high level of abstraction exceed the costs of distance from actuality. A principal benefit lies in facilitating the transition *between related formal systems*. A return on costs will be realized if we are led to alternative formalisms uniquely well-suited to stepwise refinement of the original metaphor. Minimal Chemistry Two is vindication of that strategy.

MC2 differs from the advance of MC1 over MC0 in two ways. First, in MC1 we retained the core elements of the MC0 metaphor, merely refining it to include shape. In MC2, we tinker with the ontology itself. Here we abandon λ-calculus as the chosen formalism and are empowered to do so *without* loss of progress gained in the λ framework *by virtue of an isomorphism between formalisms.* Second, unlike MC0 and MC1, MC2 has yet to be implemented. Hence, we limit ourselves below to the task of sketching, sequentially, how the typed λ-calculus leads naturally to formalisms in proof theory, how the chemical metaphor might be translated to and enriched by the proof-theoretic connection, and what limitations of a λ-based artificial chemistry this translation permits us to address.

II.3.1. From λ-calculus to proof-theory

The Curry-Howard isomorphism

Within MC1, a small number of plausible basis sets were implemented and characterized. These established that the typed system retained the capacity to yield self-maintaining organizations first established in MC0. A multitude of plausible basis sets remains unexplored. Exploring them with chemical plausibility in mind, however, requires insight into how type construction may be used to impose, for example, resource accounting *within* λ-calculus. While not impossible, the task would clearly be vastly simplified if the syntax itself imposed such a discipline of stoichiometry. Syntactical systems of this sort exist. One is led to them through a mapping between typed λ-calculus and proof-theory. Indeed, the mapping lies at the core of a deep connection between computation and logic (see [89], Appendix C.1).

Typed λ-calculus can be viewed as a syntax for the derivation of logical formulae. The mapping is known as the Curry-Howard isomorphism [38, 45] and may be stated informally as (see [89], Appendix C.1, for a more detailed introduction):

$$\begin{aligned} \text{type } \sigma &\longleftrightarrow \text{ logical formula (proposition) } \sigma \\ \lambda\text{-term of type } \sigma &\longleftrightarrow \text{ proof of } \sigma \end{aligned}$$

Since we use λ-calculus to define an abstract chemistry, any rigorous link between typed λ-calculus and other areas of mathematics extends the chemical metaphor. Roughly:

$$\begin{aligned} \text{shape} &\longleftrightarrow \text{ logical formula (proposition)} \\ \text{molecule with that shape} &\longleftrightarrow \text{ proof of that formula} \end{aligned}$$

Proof theory is a large domain, characterized by an initially bewildering diversity of syntactical systems. Our task, to which we now turn, is to situate our approach within this diversity, that is, to specify chemical interpretations of a chosen syntax which are both consistent with the isomorphism and which have potential in ameliorating the deficits of the λ-syntax.

What do proofs have to do with it?

The connection between typed λ-calculus and proof-theory and our imputed connection between both formalisms and chemistry will not demand that a reader have a rich appreciation of logic. Some prefatory remarks are, nonetheless, in order.

A logical formula is built from atomic formulae using *connectives* like `and` ($\wedge$), `or` ($\vee$), `implies` ($\rightarrow$), `negation` ($\neg$), and universal ($\forall$) and existential ($\exists$) quantifiers. There are two basic questions we may put to a logical formula. We may ask for the truth value (`true` or `false`) of a formula, or we may ask for its validity. Contrary to widespread folklore outside of logic, logic is not exhausted by "truth-tables" that permit reading off the truth-value of compound statements, such as $A \wedge B, A \vee B, A \rightarrow B, \neg A$, given the truth-values of the propositions A and B. Logic is far richer and much of the richness lies in the latter of these questions.

A taste of the relation between truth assessments and validity may be introduced by considering Frege's [30] distinction between the *sense* and *denotation* of a logical formula. In $10/2 = 1 + 4$ the denotation of both $10/2$ and $1 + 4$ is 5. Hence, the denotation of $10/2 = 1 + 4$ is `true`. How, though, does one know that $10/2$ has the same denotation as $1 + 4$? As Girard points out [36, 38], it is *not* obvious that $10/2 = 1 + 4$, for if it was we would need neither symbolism for division and addition, nor, even find need to state an equality. This is what is meant, then, by saying that $10/2$ has a different sense than $1 + 4$.

Frege's distinction tells us that what matters in logical systems is not the denotation (i.e., the content) of the propositions, but rather their relational structure. When Aristotle says that *"All men are mortal; all Athenians are men; hence all Athenians are mortal,"* he is saying *"All B is C; all A is B; hence all A is C,"* he is not saying *"true; true; hence true"* [36]. Logic manipulates the sense, not the denotation.

How then does one proceed from sense to denotation? Let us return to $10/2 = 1 + 4$. In this example, it is clear. Something must be done to show that they have the same denotation and here that "something" is a computation.

Just as different logical systems are endowed with different connectives, so must proof theory come in a diversity of flavors. The flavors of proof-theory germane to our project are those which bear a correspondence with computational operations. These comprise a class known as constructive proof theory or constructive logic. The idea behind constructive proof-theory is that the meaning of a formula is the set of its proofs, where the proofs are objects of an effective calculus, i.e., proofs are seen as construction scaffolds for a formula. Loosely speaking, the meaning of $\alpha \wedge \beta$ is to exhibit a proof of α and a proof of β. The meaning of $\alpha \rightarrow \beta$ is to provide a *function* that transforms proofs of α into proofs of β. Framed in this way, constructive logic appears not so much as a tool for reasoning about computation than a computational activity itself. Indeed, this is the essence of what is established by the Curry-Howard isomorphism.

II.3.2. Ontological commitment, resultant metaphor, and formal representation

Among the diversity of constructive logics is a recent innovation known as *linear logic* [33]. We use it here to illustrate how the proof-theoretic context refines the chemical metaphor. We do so in an intuitive fashion, building from our (similarly intuitive) characterization of typed and untyped λ-calculus. (As in our previous treatments, the reader is referred to [89], Appendix C.3, for a more detailed, but still gentle, treatment of the overall topic.)

In linear logic, indeed with proof-theory generally, the properties of a logical connective are defined algebraically (rather than by truth-tables), through rules stating how a connective can be inserted into and removed from a formula. Proof-theory emphasizes what actions must be taken to construct a formula using a set of rules governing introduction and elimination.

The introduction rules, elimination rules and connectives of a fragment of linear logic appear in [89], Appendix C.3.1. Relative to the sparsity of axiomatic operations in λ-calculus, the rules in [89], Appendix C.3.1, reveal a proliferation of syntactical operators. This meets our intent to progressively refine our chemistry (after all, one needs a richer syntax to describe an atom in quantum mechanical terms than to describe it as "a little solar system"). As the basis for the following discussion we use here only the smallest fragment of linear logic, so-called multiplicative linear logic (MLL), in sequent notation[6] ([89], Appendix C.2):

$$\frac{}{\vdash \phi, \phi^{\perp}}\ axiom \qquad \frac{\vdash \Gamma, \phi \quad \vdash \phi^{\perp}, \Delta}{\vdash \Gamma, \Delta}\ cut \tag{5}$$

Connectives

$$\frac{\vdash \Gamma, \phi \qquad \vdash \psi, \Delta}{\vdash \phi \otimes \psi, \Gamma, \Delta}\ times \qquad \frac{\vdash \Gamma, \phi, \psi}{\vdash \phi \mathbin{⅋} \psi, \Gamma}\ par \tag{6}$$

with negation, $\perp$, defined as:

$$\phi^{\perp\perp} \stackrel{\text{def}}{=} \phi \tag{7}$$

$$(\phi \otimes \psi)^{\perp} \stackrel{\text{def}}{=} \phi^{\perp} \mathbin{⅋} \psi^{\perp} \tag{8}$$

What are the imputed chemical interpretations of this syntax and how do they relate to those explored in MC0 and MC1? Recall the core of our original MC0 metaphor:

physical molecule	$\longleftrightarrow$	λ-expression (an operator)
molecule's behavior	$\longleftrightarrow$	operator's action
chemical reaction	$\longleftrightarrow$	evaluation of a functional application

Our approach in MC1 merely extended the metaphor by modulating a molecule's behavior to include:

$$\text{molecule's shape} \longleftrightarrow \text{type}$$

Composing this with the Curry-Howard isomorphism yields:

---

[6]Roughly (for readers without instant access to an appendix), $\vdash$ means "logical consequence," ψ and ϕ are propositions, capital Greek letters are sets of propositions, and $\vdash \psi, \phi$ stands for a proof of the disjunction ψ "or" (comma) ϕ using the inference rules of the proof-system. These rules are represented by a horizontal bar separating the premises (above) from the conclusion (below). An axiom is a conclusion without premises. By virtue of the **par**-rule the comma ("or") behaves like a $⅋$.

molecule's shape	$\longleftrightarrow$	type σ	$\longleftrightarrow$	logical formula σ
physical molecule	$\longleftrightarrow$	λ-term of type σ in normal form	$\longleftrightarrow$	proof of σ

The interpretation of a molecule as a proof, compelled by the isomorphism, leads to the first refinement. In linear logic (see [89], Appendix C.3), a proof is the construction of a *multi-set* of logical formulae, known as a "sequent" (whose *raison d'être* is explained in [89], Appendix C.3). Recall from Curry-Howard that a formula is equivalent to the type of a λ-expression, and that in MC1 a type was the specification of a reaction site. A proof of the sequent $\vdash \phi_1, \ldots, \phi_n$, then, represents a molecule with n possible reaction sites $\phi_1, \ldots, \phi_n$, rather than a single one as in MC1.

Now that we have logical formulae as descriptions of potential reaction sites, we need an interpretation of a chemical reaction. Like in MC1 a chemical reaction involves two issues: (i) which sites are allowed to interact (i.e., specificity), and (ii) what happens once two sites do interact (i.e., action). For this we turn to linear logic and its syntactical machinery, rules (5) and (6).

Let us again proceed intuitively. If we view a logical formula as a potentially reactive site, then we must look for the logical counterpart to "chemical complementarity" (see the discussion in Section II.2.1). In linear logic, this is provided by *negation*, $\perp$. The fact that negation is defined in terms of a duality, rules (7) and (8), fits quite naturally with chemistry, where the encounter of chemical "duals" (i.e., acid vs. base, oxidizing vs. reducing) is required for a reaction to occur.

In this logic, duality appears at the level of primitive building blocks. These are axioms, i.e., syntactical entities of the form $\overline{\vdash \psi, \psi^{\perp}}$. The structure means that a primitive object always enters the scene as a pair of connected dual "flavors," ψ and $\psi^{\perp}$. More complicated sequents (representing the reactive options of a molecule) are constructed from such "atoms" using the rules (6) for the connectives.

When two such sequents (molecules) meet, it is the cut rule that tells us whether they can interact:

$$\textbf{cut} \quad \frac{\vdash \Delta, \phi^{\perp} \qquad \vdash \phi, \Delta'}{\vdash \Delta, \Delta'} \tag{9}$$

The rule states that for an interaction to occur, each molecule must possess the *same* formula ϕ, but in a *complementary* flavor. The cut-rule, therefore, emphasizes the role of a logical proposition as an *address*, i.e., a structured name enabling communication with a specific other address (of dual kind)[7]. Girard's [34] metaphor of a *plug*, as sketched in Figure 6, renders the situation best. A proof-system, then, functions as a formal "physics" in which such addresses are constructed endogenously[8].

The cut-rule states *which* objects can interact, but it doesn't state what happens once they do so. The fact that one has a lamp and an electrical outlet does not mean one has light. All that cut does, is to *initiate* a chemical reaction (the "plugging"). An action is still required. Mathematically, the cut-rule enables two proofs to be joined into a new

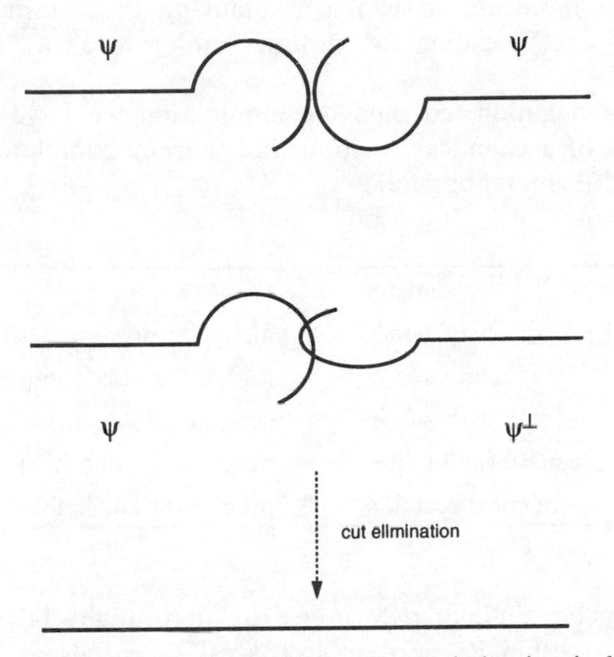

Figure 6. A cartoon of negation, cut, and cut-elimination in linear logic.

[7]In this context, the reader is urged to resist giving a logical proposition an interpretation linked to "human reasoning." It will not prove edifying. To further clarify the address issue, consider conditions under which *nothing* happens. For example, both objects carry the same flavor of sort ψ (proton-donor meets proton-donor), or both objects carry different flavors but not of the same sort (proton-donor, say ψ, meets electron-acceptor, say $\pi^{\perp}$).

[8]Chemistry may be thought of as a system for constructing addresses that enable reactions that further construct addresses. This perspective recalls the π-calculus paradigm of Milner [67–69] ([89], Appendix A.3).

proof on the basis of a "trade"; one proves (ϕ) what the other assumes ($\phi^{\perp}$). The cut rule is formally equivalent to functional application in λ-calculus which was used to initiate a chemical reaction in MC0 and MC1. Indeed, in correspondence with reduction to normal form, Gentzen [32] showed that a cut can always be removed from a proof. Crucially, Gentzen exhibited an effective process that performs this elimination by rearranging the proof structure (see [89], Appendix C.2.) It is this process, then, which is triggered by cut and that provides the associated action completing a reactive encounter. Cut-elimination yields a proof in normal form (i.e., a "direct" proof of a sequent, where the previously involved intermediate "lemma" ϕ has been removed). Completing the synthesis of a molecule is seen as "removing the intermediates" (by letting them react) yielding the product molecule as a "normal form proof."

Thus, via negation, cut and cut-elimination we have specified an interpretation of a chemical reaction and thereby completed our translation. The MC2 metaphor follows:

chemistry	···	**proof-theory**
chemical properties of bonds	···	algebraic properties of connectives
reaction site σ	···	logical formula (proposition) σ
stable molecule with site σ	···	cut-free proof of σ
chemical complementarity of sites	···	negation (σ and $\sigma^{\perp}$)
chemical reaction	···	proof with cut between σ and $\sigma^{\perp}$

This may be summarized (in words that might be chosen only by those deep within its grip) in its barest essentials as follows. The proof-theoretic metaphor views a molecule as the cut-free proof of a "shape," i.e., a chemical action. The cut-rule corresponds to a chemical reaction at site ϕ. Complementary chemical propensities are mirrored by negation, ϕ and $\phi^{\perp}$. The coming into contact of these complementary types creates an instability (the cut). In chemistry this initial site of instability is propagated through the molecular skeleton, possibly breaking old bonds and making new ones, until the stable product molecule results. This reaction progress is mirrored in the mechanics of cut-elimination that propagates the "unstable" cut-site through the proof-skeleton until normal form is attained.

II.3.3. Addressing prior limits in the linear logic framework

MC0 is an absolutely minimal chemistry, consistent with our intent in assessing the potentials inherent in adding constructive interactions to a dynamical setting. The success of MC0 in yielding self-maintaining organizations with properties seemingly so reminiscent of biological systems renders its minimality a severe impediment to progress. The deficiencies enumerated earlier (Section II.1.4) were: (i) molecules had no specificity of interaction (i.e., no shape), (ii) reactions were asymmetric by virtue of the asymmetry of application in λ-calculus, (iii) mass-conservation (stoichiometry) was violated (i.e., no resource accountability), and both (iv) reaction classes and (v) rate constants were lacking. Content that the first issue has been addressed with typed λ-calculus and progress carried over to the proof-theory framework via Curry-Howard, we here attempt to sketch the potential of the proof-theoretic representation in addressing the remaining issues. We emphasize, however, that MC2 has yet to be implemented and the following must therefore be regarded as "a transcript from a lab notebook" recorded well before the first reagents have been mixed.

Symmetry

In both MC0 and MC1 a chemical reaction was represented as functional application. Application in λ-calculus is non-commutative, hence asymmetric reactions come as a fixed deficit of the formalism. In the proof-theoretic frame a chemical reaction is modeled by the cut-rule, which is, in turn, dependent upon negation. Negation in linear logic is defined via deMorgan-like dualities (see Eq. (8) and [89], Appendix C.3) and is involutive: $A^{\perp\perp} = A$ (like a matrix transposition). This permits a completely *symmetric* reaction. By this we mean that in a reaction between sites ϕ and $\phi^{\perp}$ it doesn't matter which of the molecules carries the ϕ-site and which the $\phi^{\perp}$-site: the result is the same. Symmetry of reaction is as generic to linear logic as is asymmetry in λ-calculus.

Resource Accountability

Violation of mass-conservation is rampant within both MC0 and MC1. Recall that a failure in resource accounting obtains whenever the same variable occurs multiple times in a λ-expression. When, for example, a reaction between $\lambda \mathtt{Cl}^-.(\mathtt{Cl}^-)\lambda y.(\mathtt{Cl}^-)y$ and some $\mathtt{OH}^-$ occurs, the $\mathtt{OH}^-$ is used twice during normalization: $(\lambda \mathtt{Cl}^-.(\mathtt{Cl}^-)\lambda y.(\mathtt{Cl}^-)y)\mathtt{OH}^- \rightarrow (\mathtt{OH}^-)\lambda y.(\mathtt{OH}^-)y \rightarrow \cdots$. The meaning and extent of this difficulty is apparent. We are effectively permitting the *same* OH^--group to substitute for two distinct Cl^- ions, or, equally problematic, having the *same* Cl^- be at two different places in the molecule. Moreover, if a variable is declared that never appears in the term, a reaction would simply "annihilate" one reactant, such as in: $(\lambda \mathtt{Cl}^-.\lambda y.y)\mathtt{OH}^- \rightarrow \lambda y.y$. Serious attention to syntactical resource accounting is required in a mature artificial chemistry and no such tools are inherent in the syntax of λ-calculus.

Resource sensitivity is hardly a feature of classical logic, but several varieties of constructive logic—including linear logic—have this attribute. In classical logic, formulae are not viewed as physical entities (or tokens)—like chemicals (or money)—that are consumed when they are deployed to cause effects. A lemma proven by mathematician X need not to be reproven after mathematician Y has used it in proving a theorem. The lack of a resource problem in classical logic derives from the contraction and the weakening rules (see [89], Appendix C.2) which state that in the manipulation of proofs, available formulae can be copied or surplus instances erased arbitrarily. For example, the classical conjunction of twice the same formula, $\phi \wedge \phi$, is equivalent to ϕ. The problem with the real world is that if ϕ stands for some fact like a dollar, then classical logic states that two dollars—and, hence, any number of dollars—are equivalent to one dollar. Or that one molecule of a substance has the same effect as an Avogadro of such molecules. Here classical logic departs radically from the physical world.

Several constructive logics are resource sensitive, linear logic amongst them. Linear logic achieves resource sensitivity by placing weakening and contraction under explicit control. Depending on the tightness of the control several variant logics are obtained (e.g., [78]). The basic idea shared by each system, however strict its accounting, is to view formulae as "assets" that are consumed when they are used. For our purposes here it is sufficient to note that the syntax of linear logic permits resource accountability; attention is paid to enforce that no formula may be used that has not first been generated and that, no formula, once used, may be used again without generating it anew. This clearly permits an artificial chemistry embedded in linear logic to escape another of the deficits inherent in λ-representation.

Reaction classes

The reaction scheme of MC0 and MC1 shares a deficit beyond that of resource accounting. Specifically, the λ-framework explored only a very restricted set of chemical reaction classes. Moreover, these were exogenously imposed. The restriction is apparent in noting that a λ-expression (a reactant) applied to another can yield either a single product or no product at all. How are commonplace chemical reactions with multiple products, e.g., RCOOH + R′OH → RCOOR′ + H_2O, handled in the λ-framework? They are not.

Accommodating this deficit within the linear logic framework is the challenge that most limits implementation of MC2. In the case of λ-calculus, the limitation was inherent in the formalism. In contrast, linear logic—as well as alternative formalisms motivated by linear logic (e.g., [54])—provide a broad set of options. They raise, however, important issues of chemical interpretation (much like certain chemically motivated extensions to the logical framework raise interesting issues of logical interpretation). For the purpose of this contribution there is little to be gained in fully developing the various candidate schemes. It will suffice to introduce an example to illustrate the power of this route to MC2. We chose this particular example because it makes especially apparent a larger issue which lurks within any choice of chemical interpretation.

Recall that we identify a molecule with a proof, initiate a reaction with the inference rule "cut" and complete the reaction by cut-elimination. As with λ-calculus, it is a simple matter to generate a single product. Figure 7 illustrates a simple scheme that stretches the usual inference by yielding multiple products (i.e., more than one sequent as a conclusion). The reaction is presented in Girard's [33] proof-net notation, a concise presentation mode detailed in [89], Appendix C.3. Mastery of proof-net notation is not required to appreciate the reaction intuitively, however. It is sufficient to know that each structure is a valid proof and that the shaded regions are propositional formulae.

Imagine the figure to represent esterification of a carbonic acid with an alcohol: RCO$\boxed{\text{OH}}$ + R′O$\boxed{\text{H}}$ → RCOOR′ + $\boxed{\text{H-OH}}$ with $\boxed{\text{OH}}$ labeled (as in Figure 7) as β and $\boxed{\text{H}}$ as $\beta^{\perp}$. The cut performed on β and $\beta^{\perp}$ is now seen as the chemical action between an acid and its complementary base. The by-product of the esterification is water, captured here as the combination of the two cut-formula flavors, H and $H^{\perp} \equiv$OH, in a single molecule corresponding to $\beta - \beta^{\perp}$ in Figure 7. Clearly, exchange of shaded components between the reactant proofs has yielded the desired disconnected product proofs. The interpretation presented in Figure 7 is illustrative of how particular issues of chemical interpretation are raised. In the cut rule (9) the single conclusion sequent has lost (both flavors of) the cut formula. Yet, in our figure, we have

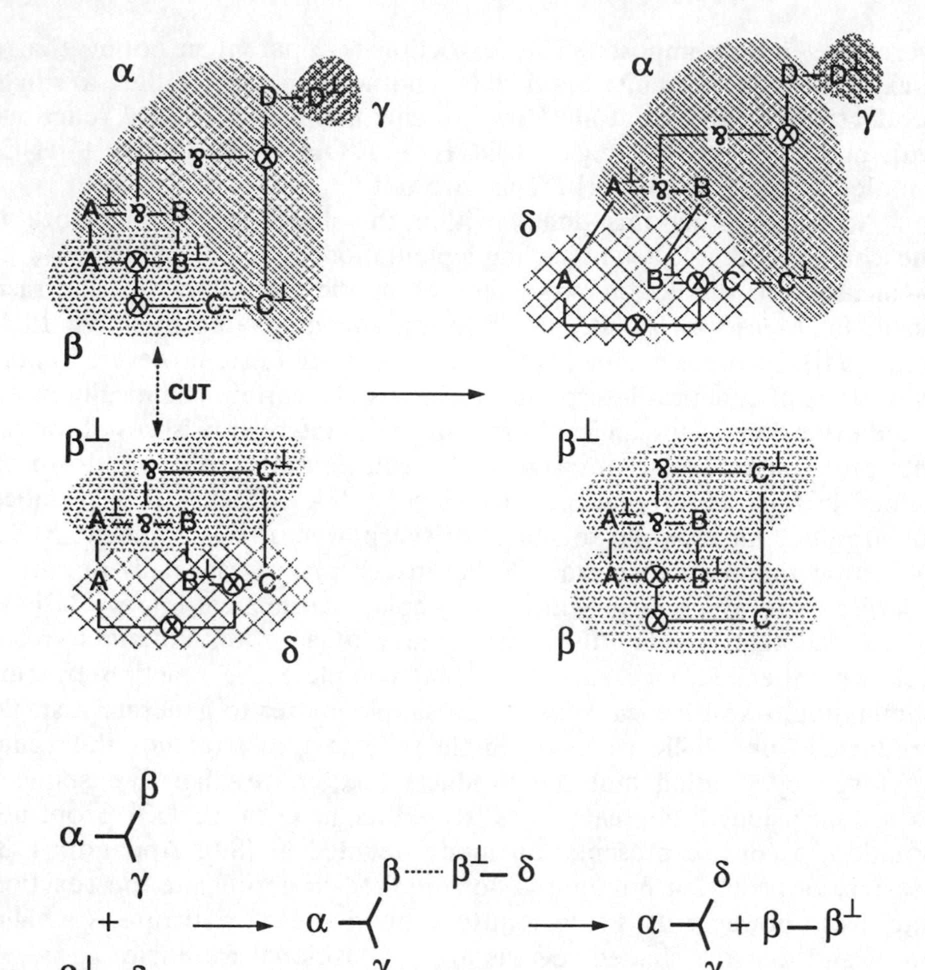

Figure 7. A chemical reaction with linear logic proof-nets. See text and [89], Appendix C.3.2 for details.

simply collected the garbage that would otherwise be thrown away during cut-elimination. One might justify this practice on the grounds that in real chemistry, nothing is ever annihilated. The attribution purchased by this justification is that of imparting a physicality to the cut-formula[9].

[9]The attribution has consequences. (i) In linear logic, the cut formula is erased. Its erasure makes cut irreversible. Keeping the latter in our interpretation (see Figure 7) preserves all information necessary for its reversal. Yet, we cannot reverse it, because the reverse reaction does not proceed via the cut

The process shuffles a reaction site from one molecule to another. This, then, is a clean interpretation of the reaction class known in chemistry as *substitution reactions.*

Treating the cut-formula as a physical object introduces a larger issue. Disarticulating the physical—whether it be an object like a molecule or a memory location in a computer—from the syntactical rules used to perform computation is a principal challenge within the domain of computer science addressing concurrency. Linear logic is one candidate formalism for concurrency. Our interpretation of the cut formula in linear logic as representing a physical object is one of several possible representations of the physical[10]. The diversity of paths within MC2 for addressing the limitation of reaction classes in MC0 and MC1 reflects this larger issue. We return to the general topic in later discussion.

Rate constants

Rate constants are an aspect of chemical reactivity with important consequences for reaction networks. In MC0 and MC1, all reactions proceed at an identical rate. A plausible foundation to endogenize rate constants

---

rule which we took as the formal definition of a reaction. (ii) Treating the cut-formula as an object, say the functional group OH, makes it impossible to subsequently cut within that object, e.g., separating the H from the O. (iii) Our interpretation only implements substitution reactions of chemistry. The scheme is unable to introduce, or eliminate, connectives (i.e., types of chemical bonds). The rules for the connectives (6) are not suited to model such reactions, because they don't provide interaction specificity (see the discussion of cut in Section II.3.2).

Other attributions to the cut formula are no less plausible, but carry their own suite of consequences. One might argue that the cut-formula has no physicality, yet retain chemical plausibility by holding that what is cut are the bonds themselves and not "that which is bound." A variant of linear logic, known as linear logic with MIX (direct logic) [2, 16, 25], readily permits both connected and disconnected proofs, and may be employed in generating multiple products. This attribution, however, purchases its power at the cost of (i) losing a strict control over weakening (and, thereby, a relaxation of strict resource accounting), as well as (ii) a far greater complexity in implementing cut-elimination.

[10] Attempting to give chemical processes a logical interpretation feeds back to logic itself. We interpret here a chemical reaction as being a logical inference (governed by a rule such as cut). But what *logical* interpretation should be given to chemical reactions that yield multiple products? That is, what is the meaning of a rule of inference that splits proofs, i.e., that generates two or more conclusion-*sequents* at the same time?

within the λ-syntax would be to interpret as a rate constant the number of reduction steps needed to obtain normal form. For this to make chemical sense, reduction must be expressed in terms of true "unit-time" events, i.e., truly elementary steps of reconfiguration. This is not the case with standard definitions of β-reduction. The linear logic framework, however, has made it possible to resolve reduction into elementary events [17, 35]. We could imagine them as taking place at the beat of an external clock. Two simultaneously occurring interactions would terminate after different periods of time, thus yielding different "rate constants."

We have focused above on the power accompanying the shift in formalism enabled by the Curry-Howard isomorphism. It is remarkable that linear logic appears to have the capacity to address *each* and to solve some of the principal limitations of MC0 and MC1. In focusing upon comparison with the λ-framework, we may have given the inadvertent impression that addressing these limitations exhausts the utility of the linear logic frame. Indeed, it may well prove to be the case that features of linear logic unrelated to the limitations of MC0 and MC1 will ultimately provide most important.

II.4. A Roadmap from chemistry to proof theory

We pause to ask: How natural is the analogy between a molecule and a proof? Suppose Dr. X claims that a certain chemical action σ is possible. Prof. Y challenges her to prove that claim. X returns a few years later and exhibits an actual molecule that provides that action, thereby proving her claim. Although the molecule does what X claimed it to do, Y will rightly wonder: "How did you produce it?". The question is fair, because by just looking at a molecule it isn't clear *how* it was produced, despite being evident *that* it was produced. The reason is that in chemistry we typically can't just stick atoms together one by one like in a Ball-&-Stick model. A molecule tells little about its synthesis path, just as a mathematician's direct proof from first principles does not convey the insight that led to it. Indeed, a mathematician usually proceeds by intermediate steps, proving lemmas, and then combining them with the cut rule (9) to achieve the desired theorem. Cut permits a proof to be factored into generic modules ("subroutines"), thereby preserving the proof idea. A mathematician, therefore, rarely normalizes the proofs.

The situation in chemistry is subtly different. First, a direct synthesis of a molecule by plugging atoms together is almost never feasible. Thus, a chemist is *forced* to synthesize a molecule by using other molecules as intermediates, i.e., he cannot fully exploit what logicians know as the subformula property. Second, as soon as the chemist mixes the reactants, the reaction proceeds, i.e., the mixture spontaneously "normalizes," courtesy of thermodynamics. The self-maintaining organizations we found with MC0 and MC1 appear from this standpoint as specific ensembles of molecules that collectively retain their synthesis pathways, because every molecule is both "final product" and "intermediate." Since in MC2 molecular actions are seen as theorems whose proofs are the molecules that perform those actions, MC2-organizations—had we produced them—would appear as sets of theorems that are closed with respect to inference. The technical word for such sets is a *theory*.

From the initial observation that dynamical systems evade construction and that computation is construction, we have arrived at a representation of chemistry as proof theory. This path has proven sufficiently surprising to us so as to expect that others might find the road map in Table 1 helpful.

The match is suggestive, and one cannot fail to wonder whether the level of metaphor might someday be trespassed. If so, there would be a level of explanation at which chemistry would effectively *be* logic. This possibility is one we do not dismiss. It is no more ludicrous than, say, becoming accustomed to regarding physical space to "be" the threefold cartesian product of the real numbers, $\mathbb{R}^3$.

III. From dynamical systems to bounded organizations: The thread from chemistry ...

The long excursion into the specific case of chemistry serves to illustrate that it is indeed possible to move beyond the limits of dynamical systems claimed at the outset, Section I. Here we return to the general point, reiterating its major features in the context both of our own attempt and those of others grappling with related issues in often quite different settings. Our treatment of related issues is admittedly eclectic, representative only of our own interests and backgrounds.

The following define our conceptual coordinates. We sequentially amplify on each point in the sections that follow.

1. The identification of the "object problem" as a fundamental limit in applying the dynamical systems methodology to the biological sciences and beyond (Section I).

Table 1. The Chemistry/Proof Theory Roadmap

Chemistry	λ-calculus	Linear logic
molecule as a physical object (the nature of atoms and their bonds)	λ-term in normal form	proof-net in normal form (multiplicative fragment of linear logic)
Shape as action (domain of interaction) (with whom and how a molecule interacts, as determined by the nature of its reactive group	type (the specification of an action: a description of what the term does at a particular level of resolution)	multiset of propositional formulae (theorem) (a set of "actionable addresses," "interfaces," or "plugging specifications")
bonds (connectors of molecular parts)	abstraction and application	left and right rules of logical connectives
initiation of a reaction	application	rule of inference: cut (reactants are the premises, products are the conclusions)
completion of a reaction (structural rearrangements into stable products)	normalization	cut elimination
branching reaction (multiple reaction pathways among the reactants)	–	multiple cut options (a sequent is a multiset)
synthesis pathway (representation of a molecule as a suite of reactions between intermediates which enable its synthesis)	a λ-term containing redexes	proof-net with cut(s) (a proof-making use of intermediate results – lemmas a.k.a cut-formulas)
synthesis planning (how do we break up a molecule into achievable subgoals for synthesis?)	theorem proving (in typed λ-calculus only, e.g., Automath [18]; see also linear logic)	theorem proving (how do we break up a formula into achievable subgoals (lemmas)?)
determinism (for a particular reactive encounter the products are determined)	Church-Rosser (reduction is confluent; in addition, some type systems have strong normalization properties, i.e., any reduction sequence always terminates)	Church-Rosser (cut elimination is confluent and obeys strong normalization in MLL)

Chemistry	λ-calculus	Linear logic
duality (dualities apparent in stereochemical complementarity, hydrophilic/hydrophobic, proton donor/acceptor, reductant/oxidant)	–	duality (linear negation is involutive, definied by DeMorgan dualities)
symmetry (reactive interaction is symmetric)	asymmetric (functional application is not commutative)	symmetric (assumption and conclusion are dual – i.e., to switch is to negate)
resource sensitive (obeys mass conservation)	–	resource sensitive (weakening and contraction are absent in MLL, or controlled in full linear logic)

2. The claim that overcoming that limit requires a theory of object construction and, thus, necessarily involves a substantive overlap with the foundations of the computational sciences.
3. The position that a concept of "organization" derives from placing a theory of objects in a suitably constrained many-body dynamical setting (Section II.1.2). The conventional settings of either dynamics alone or syntactical manipulation alone are insufficient; "organization" derives (or, if one prefers, self-organizes) from their combination in a *constructive dynamical system*.
4. Finally, the "organizations" resulting from a constructive dynamical setting have the potential to address problems that have stubbornly resisted solution.

III.1. ... to the "object problem"

The identification of the "object problem" as a fundamental limit in applying the dynamical systems methodology to the biological sciences and beyond (Section I).

Whenever a particular level of analysis of Nature is populated with objects whose internal structure engenders specific action capable of changing or creating other objects, the dynamical systems methodology encounters a fundamental limit. The reason is that the formal machinery of dynamical systems is geared to handle changes in quantities, but not changes in object structure. We believe that a formal understanding of such a level of Nature requires a theory that *combines* variables that can take objects as values with the more familiar variables that hold quantitative values, such as concentrations. To convey an intuitive flavor of this, think of action as "parametrized" by structure, and imagine a "derivative" in object space giving information about the change of object action resulting from a change in object structure. Clearly, being able to meaningfully define such a thing puts stringent conditions on how structure is coupled to action. To even start thinking about this requires a powerful formal machinery capable of expressing the coupling of structure to action for the objects pertinent to a particular domain of application. This is the "object problem."

The "object problem" is nowhere seen more crisply than in chemistry. Chemical reactions are events in which both concentrations (i.e., quantities) and objects (i.e., structures) change. The projection of a chemical reaction involving large numbers of molecules on a phase-space of concentrations is known as reaction kinetics. To set up a chemical reaction as a dynamical system in concentration space, one only requires knowledge of the proper couplings among the concentrations of reactants and products. It is sufficient if these are known as empirical facts; knowledge of the chemical identity and properties of reactants and products is not necessary. Cranking the tools of, say, infinitesimal calculus yields the time evolution of reactant and product concentrations. Remove kinetics for a moment by considering just the information conveyed by a chemical reaction when it is notated on paper. We are left with a reaction arrow, "$\rightarrow$", expressing a *relation* among molecular structures. A general method capable of describing the time evolution of the contents of a reaction vessel for an *arbitrary* initial mixture of molecular species would require nothing less than a formal system implicitly representing the space of molecular objects and the relation "$\rightarrow$" over them. That is a formal theory of chemistry.

III.2. . . . to the foundations of mathematics

The claim that overcoming that limit requires a theory of object construction and, thus, necessarily involves a substantive overlap with the foundations of the computational sciences.

Where do we get a formal theory of chemistry? The answer is crucial to our approach. To date we do not have a formal, axiomatized theory of chemistry that is useful in everyday practice, despite the fact that quantum mechanics successfully grounds chemistry in the behavior of electrons and nuclei. The problem is one of choosing the "right" level of description. With respect to both molecular biology and industrial metabolisms alike, quantum mechanics is far too fine grained, and, aside from issues of feasibility, does not convey a satisfactory understanding of "what chemistry is actually doing"; it is too close to the trees to see the forest. To put it provokingly, our understanding of life will derive in large measure from *how* we understand chemistry. It is clear, then, that identifying a coarser grain of analysis capable of hosting a formal theory of chemistry would be of tremendous practical import and by no means limited to the foundations of theoretical biology.

Chemistry and computation

The stance we took in prior work [26, 27], and further elaborated here, is based on the intuition that at some level of description the reactive processes of chemistry are *analogous* to manipulations (rewrites) of syntactical objects. This puts us right into the domain of the computational sciences, Section II.1.1. In the present context, the reader is well advised to detach from an all-too narrow notion of computation as "number crunching." Much effort in the computational sciences goes into devising formal systems of syntactical constructs (we call "objects") that are interrelated by operations of transformation defined on them. It is in this sense that "computation" is the science of the construction of abstract objects with structure-specific "behavior."

What is crucial in the present context is *how* object behavior is synthesized from basic elements, as it is here that insights into fundamental mechanisms of "construction" or "interaction" are revealed, and can be compared with empirical facts. For what is desired in a theory of objects is not just a formalism and the theorems that accompany it, but a transparency of interpretation in the intended application domain. In a chemical application, one wants to capture *at least* the twin facts that (i) product molecules are lawfully constructed from substitution of parts of reactant molecules and (ii) that the same product can be produced by a diversity of different reactants. One wants, therefore, a theory of *combinational structures* and *substitution* together with the resultant theory of *equality*. Indeed, this is what drew us originally to λ-calculus. A great variety of alternative formal systems—Turing machines, Petri nets, Post systems, cellular automata, to mention but a few—allow expression of the same set of functions on the natural numbers, and, thus, may be regarded as being equivalent *in that respect*. It should be clear by now, however, that what is needed here is not merely a member of this universality class, but a member whose features are germane to the chemical problem at hand.

Identifying λ-calculus as a plausible candidate for a chemical interpretation hardly qualifies λ-expressions as anything but the most metaphorical of molecules. It is merely a foothold. A crucial one, however. It enabled our painstaking progression from MC0 to MC2, showing how a formal representation can be preserved, while progressively refining the chemical interpretation of the operators of the formalism. This would have hardly been possible with, say, Turing machines, Boolean operators, or assembler code. In that respect we differ markedly from pertinent work by Kauffman [47, 49], Rasmussen [76], McCaskill [62], Thürk [85] and Ikegami [46].

The issue here is one of grounding and formalizing an ontology, not just of capturing a phenomenology. Yet, this does not require a one-to-one mapping between some formalism expressing computational processes and real-world molecules with their chemical reactions. The map should not be confused with the territory; we do not want to "simulate" chemistry. We take the computational perspective as one enabling a different—logical—level of description of chemistry which is distinct from one that accounts for its actual physical implementation. The latter is the domain of quantum physics. For example, whether the actual protein folding process belongs to the class of computable functions is entirely irrelevant to us. For all we need to capture is the *logic* of the connection between structure and action specificity, not the physical process by which this connection is implemented. Herein lies the point of a specification language for chemistry.

What is gained by a computational theory of objects?

What is gained is best seen in comparison with prior approaches to modeling chemical collectives [3, 24, 48, 70, 79], following Kauffman's [47, 48] original casting of the problem. Particularly germane here are the efforts of Bagley, Farmer, Kauffman and Packard [3, 24]. Their work is an ideal contrast, in that the objective is identical to our own, but the methodology used to achieve the end differs. In its essence their model consists of strings over some alphabet, for example 0s and 1s, meant to represent polymers that recognize each other by "complementarity" (0 pairs with 1). A string can act as the docking place for others, thereby catalyzing specific concatenation and splicing reactions. This leads to the assembly of reaction networks capable of maintaining themselves on the basis of a monomer or string flow through the system. Such a model shares with our own the appearance of self-maintaining collectives (that they call autocatalytic).

The crucial difference between our approaches is that the core of our model is a *theory* of object construction—rather than the imitation of particular chemicals. This is what gives us the capacity to specify what the "organization" of an emerging collective of objects is in terms of a mathematical formalism. Three broad consequences follow.

1. The abstraction of molecules in MC0 and MC1 as symbolic functions allows organization to be detected as a closure of interaction, manifested by invariant syntactical regularities and invariant algebraic laws characterizing the action of those objects maintained in the collective. It cannot be overemphasized that this characterization can be made by an observer of the system who is ignorant of λ-calculus. Indeed, an organization can be specified as an algebraic rewrite system that is independent from λ-calculus, and, thus, the process by which it originated. With the theory of objects comes whole-cloth a quite different theory of the collective.
2. A formal theory of objects makes transparent which features of the collective organization derive from the underlying theory of objects, and which features are curiosities derived from particular initial conditions, parameter settings, or a particular chemical stance. The distinction is between what we have elsewhere called "digital naturalism" [29] and the claim for a theory of self-maintenance.
3. As emphasized at the outset, Section I, an abstract theory of objects plays a role analogous to that of differential equations. The analysis of a dynamical system cast in terms of differential equations yields the characterization of manifolds in phase space that govern the set of its *possible* trajectories. Consider our MC0 implementation and imagine we seeded our reactor with one λ-expression known to be a basis for all λ-calculus. Imagine further that the container is itself infinite in size and the reactor would then be capable of holding all possible (normal form) expressions. When we impose a dynamics on the objects (which, in our case, is a scheme coupled to the reactor size), we sieve particular "trajectories" in object space. Recall that our kinetic scheme is designed to favor the maintenance of objects that are constructed by the extant population of objects. As a

consequence, the "trajectory" in object space "converges" to an "attractor"—a self-maintaining organization.

This casting emphasizes the need for a theory of the "motion" in "object space" induced by the object constructors (here, functional application or logical inference) under the continuously updating kinetics imposed by the extant network of objects. How might such a motion differ under dynamics different from our own (see for example [46])? Or with equivalent dynamics and different object-constructors? Is there a meaningful formal concept of a "trajectory" in "object space"? What is "continuity"? Is there a useful definition of "distance" between "attractors" (in our case, algebraic structures in λ-space)? Questions like these require a theory of objects, not only to be answered, but even to be asked.

The value is apparent. Consider just one instance. A methodological imperative of the dynamical systems approach is the *a priori* choice of the pertinent entities and their functional couplings defining the system. The fact that the choice must be made *a priori* has the consequence that *the dynamical systems methodology can never be used to address the origin of that same system*, a profound limitation to which we have referred elsewhere ([26]) as the existence problem. It is apparent in MC0 that our reactor in settling upon a self-maintaining set of objects has settled upon a fixed system of variables and functional couplings between them; thus, *particular dynamical systems appear as limiting cases (such as "fixed points") of constructive dynamical systems.* Are constructive dynamical systems "generators" of dynamical systems? If so, a formalization of the motion in such a space holds promise as a methodology to address scientific questions which include the phrase "the origin of . . . ."

Grounding and unifying other's ideas

Attractive intellectual constructs previously lacking a formal interpretation are rendered accessible to conventional modes of scientific investigation in our setting. Maturana and Varela's concept of "autopoiesis" [56–58, 87, 88] is particularly close, indeed arguably indistinguishable, from our concept of organization. It shares the key ingredient that the system is composed of "components" which engage in a network of interactions that enable the continuous regeneration of these same "components." Thus the autopoietic system is, at essence, a matter of constructive relationships closed upon interaction; this Varela labels autonomy (we say, self-maintenance). Autopoietic systems share with our

λ-organizations a number of other features, including regenerative abilities, accessibility only to inputs that influence "component" interrelations, and capacity for hierarchical coupling.

Rosen's (M,R)-system (for metabolism-repair system) [20, 80] resembles an autopoietic system, but Rosen's "components" are pure abstract functionalities. Rosen packs a whole metabolism into a single functional letter (a "metabolic function"), or speaks of a "repair function" and a "replication function," the three of which entail each other in a circular fashion by mutually acting on their domains and ranges. The point that we find concordant here is Rosen's emphasis on the causal circularity inherent to functional organization. Note, however, that Rosen's (M,R)-systems lack any notion of object construction. Rosen claims that (M,R)-systems are inherently unformalizable. Casti [11, chapter 7] approaches some of their aspects by means of dynamical systems.

Thus both constructs—autopoietic systems and (M,R)-systems—share with ours the essential notion of closed relations of construction between the parts of the system. Our work departs from both, however, in providing a concrete theory of the conditions necessary to realize a universe of such systems and to characterize their features in a standard formal setting.

We suspect that the autopoietic concept differs *only* as a consequence of Maturana and Varela and subsequent investigators [71] having come to it without the benefits of viewing organization as the consequence of joining dynamics and construction. The only claim of Maturana and Varela that is not instantiated in our organizations is their requirement that the system be spatially bounded. This is essential for them, for it is the only device by which their "components" may be isolated from the "rest-of-the-world." The seeming need of a membrane laid out in space is, in our view, only required because the characterization of autopoietic systems is not built upon a theory of its components. Our organization are indeed bounded, but bounded syntactically (i.e., λ-organizations are special invariant subspaces of λ-space). A bounding is indeed a necessary feature of organizations, but the space need not be 3-space. Perhaps it is not surprising that several disciplines which have found the concept of autopoiesis of utility (e.g., notably law and social psychiatry) find the requirement of spatial bounding dispensable (see review by [71]). At minimum, then, our work has converged to a notion similar to that of autopoiesis from an independent angle; quite plausibly, though, we have unwittingly generated a formal interpretation of a heretofore frustratingly elusive notion of considerable importance.

III.3. ... to concurrency and self-organization

The position that a concept of "organization" derives from placing a theory of objects in a suitably constrained many-body dynamical setting (Section II.1.2). The conventional settings of either dynamics alone or syntactical manipulation alone are insufficient; "organization" derives (or, if one prefers, self-organizes) from their combination in a constructive dynamical system.

It bears emphasis that "objects," as we frame them here, are defined by structure-action relationships *where each action is a mapping from structures to structures*. In a many-body setting this generates a *constructive feed-back loop* (in analogy to dynamical feed-backs) which causes the emergence of "organization." The so-defined constructive feed-back is absent in genetic algorithms [39, 44], genetic programming [53], classifier systems [43], and models of evolutionary optimization [1, 28]. While these systems deal with objects whose structure entails action, the action does not participate in object construction. This is exactly what puts our concept of organization outside their scope.

We hesitate to attribute to constructive dynamical systems claims of "emergence" or "self-organization," in that these terms are increasingly used with quite different attributions. Our organizations, however, do emerge in the sense that an organization possesses (*ex post*) a level of description that is independent of the abstract chemical universe within which it originated. Similarly, the core-objects (constructors) of the organization self-organized in the restricted (but meaningful) sense that the constructive dynamical system converged to them by an endogenous motion in object space.

Our use of the words self-organization and emergence differ sharply from the frequent use of these terms as meaning "a phenomenon displayed by a collective and unexpected by the investigator." The distinction is, again, one between a theory of the collective grounded in a formalism to which an interpretation (a meaning) is given to the operators of the formalism and the observation of an instance of collective phenomena for which no underlying theory guides interpretation or guarantees generality (i.e., "digital naturalism" *sensu* [29]). In drawing this distinction, we intend no disrespect for the value of such "complex systems" studies. In at least two cases a lack of formal grounding of components of the collective is appropriate. First, it is appropriate whenever the underlying suite of behaviors of the components are themselves empirically established to be disassociable from the features of the system left unmodeled. Examples include much of individual-based modeling in behavioral and community ecology (i.e., there is no need for a theory of molecules-as-proofs to study the consequences of odor trails on patterns of ant dispersion). Second, it is often desirable to leave uninterpreted the nature of the objects when one is seeking to implement a system whose objective is a search. Holland's genetic algorithms [44] and classifier systems [43] need not be faithful chromosomes or genotype-to-phenotype representations when the intended objective is an efficient search engine.

A more appropriate embedding of constructive dynamical systems lies in the domain of computer science addressing concurrency. The ground here is not so much infertile as it poorly prepared for sowing. Susan Oyama's underground classic, *The Ontogeny of Information* [73], documents the lack of rudimentary intellectual hygiene in the free use of metaphor from computer science to describe and interpret biological observations. The attributions are ubiquitous: the genes as "code," the genome as "algorithm," the cell as "massively parallel computer," and the like. Indeed, the metaphors are not merely passive inaccuracies in the service of rhetorical aims; they actively frame our thinking (see, for example, [74] on representations of the *ras* pathway). We will not tread lightly here. *The use of the computational metaphor is crucially vacuous without a formal translation between a chemical syntax and a syntax of computation.* Here we reawaken the problem of interpretation; our as-yet-incomplete march from MC0 to MC*n* has achieving this formal interpretation as its intent.

It is, indeed, our stance that the formalisms *natural* to biology derive not from physics (as a discipline), but from imposing those of physics on those of computer science. Organisms are coherent chemical collectives, molecules are constructed from molecules, and computation is a science of representing possible constructions. The control of timing the interaction of physical objects and the representation of the same objects as performing constructions (computations) lies at the heart of both realizing a genuinely parallel computer and our attempt to develop a meaningful representation of a chemical parallelism (viz our discussion of reaction classes in MC2, Section II.3.2). Disarticulating the physical from the computational is what makes parallelism a hard problem in computer science.

The relation is sufficiently subtle to remind the reader of the manner in which our organizations were realized. The strategy we adopted was two-pronged: we first have projected chemical objects onto abstract logical entities, using λ-calculus as a formalization of their constructive interactions. Then we have dipped these entities into a cocktail of chosen physical modalities: (i) many objects *coexist* in the same system (reactor), (ii) an object-species has a *concentration* (objects can occur in multiple instances), (iii) objects interact by *random collision*, (iv) objects have a *finite lifetime* (constrained flow reactor).

Point (i) enables a constructive feed-back loop by providing a *context* from which the interaction partners of an object are drawn, and to which the product of a reaction is returned. Points (ii) and (iii) provide a simple (pseudo-chemical) kinetics that biases interactions to occur among object species with the highest concentration (*relevance*). Point (iv) implements the overall effect of removal, such as loss, decay, or inactivation. In conjunction with points (ii) and (iii) this removal reinforces precisely those reaction pathways whose constituents on the reactant side also appear on the product side somewhere along the path. The result are self-maintaining collectives of objects, with each object *being simultaneously interpreted as a physical object and a function.*

We are drawn to this view of parallelism by physical intuition—flow reactors and chemicals being common laboratory objects. The computer scientist comes from a quite different intuitional base. Physicality is foreign ground; the strength of computation by syntactical manipulation lies in its leaving open the interpretation of the syntax. We find it remarkable, then, to what degree our work abrades with that of theoretical computer science. The body of work known as *communication and concurrency* [42, 64, 65, 67], aims at a formalization of the behavior of systems consisting of many coexisting and independently interacting heterogeneous computational agents. Individual agents are referred to as "processes." Processes "communicate" to influence each other's behavior (i.e., the ability to communicate). Two examples serve to illustrate how short the intellectual distance is separating the issues in concurrency and our approach to organization.

In π-calculus [68, 69], a popular concurrent formalism ([89], Appendix A.3), a single expression is equivalent to the entire content of our reactor at a given point in time. The evolution of the reactor appears as a series of "reductions" of this large expression. Even more striking is the correspondence between our system and a device used to choose among the multiple communication channels available to a given concurrent process. To address this issue, Berry and Boudol [5, 6] introduced the notion of a Chemical Abstract Machine as a possible execution machine for the π-calculus. Berry and Boudol's insight—which predates our work—was to use physical aspects of chemistry (such as randomly colliding objects) to implement a concurrent computation, while we independently originated the reverse interpretation, i.e., to use computational objects as a proxy for molecules and to join them together into a concurrent setting. Despite the distance in intended application, then, both approaches share the underlying challenge of imposing a physics upon computational construction. Computer scientists seeking to *implement* concurrency no longer have the luxury of ignoring physical modalities—their processes must communicate in time to other processes with "real-world" positions and properties. This is, again, little different from our own problem in reverse; we start from a dynamical systems setting and need to add to it a formalism of object construction. We both face the challenge of confronting time and properties that characterize objects as physical entities, while simultaneously endowing the same entities with the power of construction.

Despite this correspondence, our efforts depart sharply from work in concurrency and communication in one important respect. Our organizations self-organize from random communications in a many-body, flow-reactor setting. Self-organization is anathema to computer engineers—indeed, such lack of rigid control over communication is precisely what they seek a formalism to avoid. Control over communication is deemed essential; one need only imagine the task of a systems administrator attempting to communicate with an operating system after a series of system calls have unexpectedly self-organized. Yet, as our organizations indicate, the product of constructive dynamical systems is not a lack of coherent behavior, but the creation of collectives with predictable features and properties. Perhaps the treatment of concurrency would benefit from exploring the feature in which our work departs from their own. Surely—at the moment—the prospect of e-mail messages spontaneously combining into a coherent manuscript strikes the authors as an acceptable price to pay for an occasional unexpected core dump!

III.4. ... to biology and beyond

Finally, the "organizations" resulting from a constructive dynamical setting have the potential to address problems that have stubbornly resisted solution.

Biology has only two claims to theories unto itself—Mendel's theory of transmission and Darwin's theory of natural selection. The intellectual history of the first half of this century is a story of continuing debate over whether the two theories were in conflict. Fisher, Haldane, and Wright demonstrated that no conflict existed and the same fields are filled today by a self-perpetuating army of investigators using the *same plows* (powered now by computers much as farmers today use tractors). The talk of plows and tractors is not intended as idle ridicule, for the tools are the issue here; the limits of the tools define the "barrier" we address. The tools employed by Fisher to show that the great theories of biology were concordant required casting the problem in a fashion that threw out the constructive aspects of biology, rendering the problem tractable as one in dynamical systems. Throwing out construction meant throwing out the organism; trying to put the organism back in is fair epitome of the intellectual history ever since.

The claim is that biology requires a trinity of theories; we have two of them; we lack only "a theory of the organism." The claim we make is that the self-maintaining organizations we derive hold promise as that missing theory. Indeed, such a theory need not await a global solution to the specification language for chemistry. The fact that our organizations can be described in a formalism distinct from that in which they were generated (i.e., as abstract rewrite systems rather than λ-expressions) leaves their terms (like those of any syntax) open to interpretations other than chemical. From this realization flows a diversity of potential applications.

Given a universe of self-maintaining abstract rewrite systems (ARS), the uses are limited solely by the properties of the particular ARS and the interpretation given to its terms. An interpretation of the terms as engineering functions in a machine might be route to a self-repair mechanism, an interpretation as a semiotic unit as a device for natural language interpretation, and interpretation of terms-as-molecules as germane to a blueprint for the design of a self-maintaining chemical manufacturing process [7], as it is to a metabolic cycle in a cell or a system of cell-cell communications defining an organ. The origin of such specifications from a research program in artificial chemistry or in experimental λ-calculus is irrelevant. It cannot be overemphasized that herein lies the significance of the characterization of our organizations in an alternative formalism.

To the extent that one accepts that the missing "theory of the organism" is recognizable as a general specification procedure for self-maintaining systems of constructive relations, MC0 suffices. Application to the biological issues left wanting for a generation—indeed to domains distinct from biology—are limited solely by the interpretation given to the terms of the abstract rewrite system and the extent to which its properties are germane to the question at hand. While work-in-progress portends considerable promise in applications to evolutionary biology in particular, it is neither feasible nor appropriate to address them here. After all, the editors asked us to identify a "barrier to knowledge," they did not ask us to lift it.

The claim for relevance here is large indeed. Hence, we conclude with a warning. The optimism and the passion with which we assess the potential of constructive dynamical systems has all the characteristics of a bullish investor at the eve of the market's collapse. The reader would do well to heed the admonition

... beware of the boa constructor.
Erwin Panofsky

Acknowledgements: We thank Harald Freund for numerous discussions on the limits and potentials of MC0. This is paper #41 from the Center for Computational Ecology at Yale University.

References

[1] Amitrano, C., L. Peliti and M. Saber. Population dynamics in a spin-glass model of chemical evolution. *J. Molec. Evol.*, 1990, 29, 513–525.

[2] Asperti, A. Causal dependencies in multiplicative linear logic with MIX. *Math. Struct. in Comp. Sci.*, 11, 1993, 1–31.

[3] Bagley, R. J. and J. D. Farmer. Spontaneous emergence of a metabolism. C. G. Langton, C. Taylor, J. D. Farmer and S. Rasmussen, eds. *Artificial Life II.* Santa Fe Institute Studies in the Sciences of Complexity, 93–141, 1992, Redwood City, CA., Addison-Wesley.

[4] Barendregt, H. G. *The Lambda Calculus: Its Syntax and Semantics.* 2d revised ed. Studies in Logic and the Foundations of Mathematics, North-Holland, Amsterdam, 1984.

[5] Berry, G. and G. Boudol. The Chemical Abstract Machine. *17th ACM Annual Symposium on Principles of Programming Languages.* 1990, New York, ACM Press, 81–94.

[6] Berry, G. and G. Boudol. The Chemical Abstract Machine. *Theoretical Computer Science,* 1992, 96, 217–248.

[7] Bro, P. *Artificial life in real chemical reaction systems.* Book manuscript in preparation (contact address: P. Bro, Santa Fe Institute, 1399 Hyde Park Road, Santa Fe NM 87501).

[8] Buss, L. W. *The Evolution of Individuality.* Princeton University Press, Princeton, NJ, 1987.

[9] Cardelli, L. Type systems. Chapter in a forthcoming *CRC Handbook of Computer Science and Engineering,* available at `http://www.research.digital.com/SRC/personal/Luca_Cardelli/Papers.html`. 1996.

[10] Cardelli, L. and P. Wegner. On understanding types, data abstraction, and polymorphism. *Computing Surveys,* 17, 1985, 471–522.

[11] Casti, J. *Reality Rules–II.* Wiley, New York, 1992.

[12] Church, A. A set of postulates for the foundation of logic. *Annals of Math. (2)*, 1932, 33, 346–366.

[13] Church, A. A set of postulates for the foundation of logic (Erratum), *Annals of Math. (2)*, 1932, 34, 839–864.

[14] Church, A. *The Calculi of Lambda Conversion.* Princeton University Press, Princeton, 1941.

[15] Damas, L. and R. Milner. Principal type-schemes for functional programs. *Proceedings of the 9th Annual Symposium on Principles of Programming Languages,* 207–212, 1982, New York, Association of Computing Machinery.

[16] Danos, V. and L. Regnier. The structure of the multiplicatives. *Arch. Math. Logic,* 28, 1989, 181–203.

[17] Danos, V. and L. Regnier. Proof-nets and the Hilbert space. In J.-Y. Girard and Y. Lafont and L. Regnier, eds. *Advances in Linear Logic.* London Mathematical Society Lecture Note Series, 307–328, 1995, Cambridge, Cambridge University Press.

[18] de Bruijn, N. G. A survey of the project Automath. In J. R. Hindley and J. P. Seldin, eds. *To H. B. Curry: Essays on Combinatory Logic, Lambda-Calculus and Formalism.* Academic Press, New York, 579–607, 1980.

[19] Szabo, M. E., ed. *The Collected Papers of Gerhard Gentzen.* North Holland, Amsterdam, 1969.

[20] Rosen, R., ed. *Foundations of Mathematical Biology,* (vol.2). Academic Press, New York, 1972.

[21] Eigen, M. Self-organization of matter and the evolution of biological macromolecules. *Naturwissenschaften,* 1971, 58, 465–526.

[22] Eigen, M., J. S. McCaskill and P. Schuster. The Molecular Quasi-Species. *Advances in Chem. Phys.,* 1989, 75, 149–263.

[23] Eigen, M. and P. Schuster. *The Hypercycle.* Springer Verlag, Berlin, 1979.

[24] Farmer, J. D., S. A. Kauffman and N. H. Packard. Autocatalytic replication of polymers. *Physica D,* 1982, 22, 50–67.

[25] Fleury, A. and C. Retoré. The mix rule. *Math. Struct. in Comp. Sci.,* 4, 1994, 273–285.

[26] Fontana, W. and L. W. Buss. The arrival of the fittest: Toward a theory of biological organization. *Bull. Math. Biol.,* 1994, 56, 1–64.

[27] Fontana, W. and L. W. Buss. What would be conserved 'if the tape were played twice'? *Proc. Natl. Acad. Sci. USA,* 1994, 91, 757–761.

[28] Fontana, W., W. Schnabl and P. Schuster. Physical Aspects of Evolutionary Optimization and Adaptation. *Phys. Rev. A,* 1989, 40, 3301–3321.

[29] Fontana, W., G. Wagner and L. W. Buss. Beyond digital naturalism. *Artificial Life*, 1, 1994, 211–227.

[30] Frege, G. *Begriffsschrift, eine der arithmetischen nachgebildete Formelsprache des reinen Denkens.* Louis Nebert, Halle, 1879.

[31] Freund, H. Self-maintaining λ-organizations and analysis via rewrite systems. Poster presented at the 3rd European Conference on Artificial Life (ECAL 95) in Granada, Spain, June 1995.

[32] Gentzen, G. Untersuchungen über das logische Schließen. *Mathematische Zeitschrift,* 1935, 39, 176–210, and 405–431.

[33] Girard, J.-Y. Linear logic. *Theoretical Computer Science,* 50, 1–102, 1987.

[34] Girard, J.-Y. Towards a geometry of interaction. In J. W. Gray and A. Scedrov, eds. *Categories in Computer Science and Logic,* pp. 69–108. American Mathematical Society, 1989. Proceedings of the AMS-IMS-SIAM Joint Summer Research Conference, June 14–20, 1987, Boulder, CO; Contemporary Mathematics, Vol. 92.

[35] Girard, J.-Y. Geometry of interaction II: Deadlock-free algorithms. In P. Martin-Löf and G. Mints, eds. *COLOG-88,* pp. 76–93. Springer-Verlag, LNCS 417, 1990.

[36] Girard, J.-Y. La Logique linéaire. *Pour La Science, Edition Française de 'Scientific American',* 150, 74–85, April 1990.

[37] Girard, J.-Y. Linear logic: Its syntax and semantics. In J.-Y. Girard, Y. Lafont and L. Regnier, eds. *Advances in Linear Logic,* pp. 1–42. Cambridge University Press, 1995. Proceedings of the Workshop on Linear Logic, Ithaca, NY, June 1993.

[38] Girard, J.-Y., Y. Lafont and P. Taylor. *Proofs and Types.* Cambridge

Tracts in Theoretical Computer Science 7, Cambridge University Press, 1988.

[39] Goldberg, D. E. *Genetic Algorithms in Search, Optimization and Machine Learning.* Addison-Wesley, Reading, MA, 1989.

[40] Hankin, C. *Lambda Calculi. A Guide for Computer Scientists.* Clarendon Press, Oxford, 1994.

[41] Hennessy, M. *Algebraic Theory of Processes.* MIT Press, Cambridge, MA, 1988.

[42] Hoare, C. A. R. *Communicating Sequential Processes.* Prentice Hall, Englewood Cliffs, 1985.

[43] Holland, J. Escaping brittleness: The possibilities of general purpose machine learning algorithms applied to parallel rule-based systems. In R. S. Michalski, J. G. Carbonell and T. M. Mitchell, eds. *Machine learning: An artificial intelligence approach.* Kaufmann, Los Altos, CA, 1986.

[44] Holland, J. H. *Adaptation in Natural and Artificial Systems.* Bradford Books, MIT Press, Cambridge, MA, (reprint edition), 1992.

[45] Howard, W. A. The formulae-as-types notion of construction. In *To H. B. Curry: Essays on Combinatory Logic, Lambda-Calculus and Formalism.* J. R. Hindley and J. P. Seldin, eds., 479–490, Academic Press, New York, 1980.

[46] Ikegami, T. and T. Hashimoto. Coevolution of machines and tapes. In F. Morán, A. Moreno, J. J. Merelo and P. Chacón, eds. *Advances in Artificial Life.* Third European Conference on Artificial Life, Granada, Spain, 234–245, 1995, Berlin, Springer Verlag.

[47] Kauffman, S. A. Cellular homeostasis, epigenesis and replication in randomly aggregated macromolecular systems. *J. Cybernetics,* 1971, 1, 71–96.

[48] Kauffman, S. A. Autocatalytic sets of proteins. *J. Theor. Biol.,* 1986, 119, 1–24.

[49] Kauffman, S. A. *The origins of order.* Oxford University Press, New York, 1993.

[50] Klop, J. W. Term Rewriting Systems. In S. Abramsky, D. M. Gabbay and T. S. E. Maibaum, eds. *Handbook of Logic in Computer Science.* 2, 1–116, Clarendon Press, Oxford, 1992.

[51] Klop, J. W. and A. Middeldorp. An introduction to Knuth-Bendix completion. *CWI Quarterly,* 1, 1988.

[52] Knuth, D. E. and P. E. Bendix. Simple word problems in universal algebra. In J. Leech, ed. *Computational Problems in Abstract Algebra.* Pergamon Press, New York, 1970.

[53] Koza, J. R. *Genetic Programming: On the Programming of Computers by Means of Natural Selection.* MIT Press, Cambridge, MA, 1992.

[54] Lafont, Y. From proof-nets to interaction nets. In J.-Y. Girard, Y. Lafont and L. Regnier, eds. *Advances in Linear Logic.* London Mathematical Society Lecture Note Series, 225–247, 1995, Cambridge, Cambridge University Press.

[55] Lalement, R. *Computation as Logic.* Prentice Hall, Englewood Cliffs, NJ, 1993.

[56] Luisi, P.-L. Defining the transition to life: self-replicating bounded structures and chemical autopoiesis. In W. Stein and F. J. Varela, eds. *Thinking About Biology.* Santa Fe Institute Studies in the Sciences of Complexity, 3–23, 1993, Redwood City, CA, Addison-Wesley.

[57] Maturana, H. and F. J. Varela. *De Máquinas y Seres Vivos: Una teoría de la organizacíon biológica.* Editorial Universitaria, Santiago, Chile, 1973. (Reprinted in: H. Maturana and F. J. Varela, *Autopoiesis and Cognition: The Realization of the Living*, 1980.)

[58] Maturana, H. and F. J. Varela. *Autopoiesis and Cognition: The Realization of the Living.* Reidel, Boston, 1980.

[59] Mauny, M. Functional programming using Caml Light. (User's manual available from ftp.inria.fr by anonymous ftp.) January 1995.

[60] Maynard Smith, J. Natural selection and the concept of a protein space. *Nature,* 1970, 255, 563–564.

[61] Maynard Smith, J. and E. Szathmáry. *The major transitions in evolution.* Freeman, Oxford, 1995.

[62] McCaskill, J. S. Polymer Chemistry on Tape: a computational model for emergent genetics. Unpublished manuscript, MPI für biophysikalische Chemie, Göttingen, 1988.

[63] Milner, R. A theory of type polymorphism in programming. *Journal of Computer and System Sciences,* 1978, 17, 348–375.

[64] Milner, R. *A Calculus of Communicating Systems.* Lecture Notes in Computer Science, Vol. 92, Springer-Verlag, Berlin, 1980.

[65] Milner, R. *Communication and Concurrency.* Prentice Hall, Englewood Cliffs, NJ, 1989.

[66] Milner, R. The polyadic π-calculus: a tutorial. Report ECS-LFCS-91-180, University of Edinburgh, 1991.

[67] Milner, R. Elements of interaction. *Comm. ACM,* 1993, 36, 78–89.

[68] Milner, R. J. Parrow and D. Walker. A Calculus of Mobile Processes, I. *Information and Computation,* 1992, 100, 1–40.

[69] Milner, R., J. Parrow and D. Walker. A Calculus of Mobile Processes, II. *Information and Computation,* 1992, 100, 41–77.

[70] Minch, E. Representation of Hierarchical Structure in Evolving Networks. State University of New York at Binghamton, PhD Dissertation, 1988.

[71] Mingers, J. *Self-Producing Systems – Implications and Applications of Autopoiesis.* Plenum Press, New York, 1995.

[72] Morowitz, H. J. *Beginnings of Cellular Life: Metabolism Recapitulates Biogenesis.* Yale University Press, New Haven, CT, 1992.

[73] Oyama, S. *The Ontogeny of Information.* Cambridge University Press, Cambridge, 1985.

[74] Oyama, S. The accidental chordate: contingency in developmental systems. In B. H. Smith and A. Plotnitsky, eds. *Mathematics, Science, and Postclassical Theory (South Atlantic Quart., Vol. 94),* 509–526, Duke University Press, Durham, NC, 1995.

[75] Pratt, V. R. The duality of time and information. In *Proceedings of the Third International Conference on Concurrent Theory,* W. Cleaveland, ed., 237–253, 1992, New York, Springer-Verlag.

[76] Rasmussen, S., C. Knudsen, R. Feldberg and M. Hindsholm. The coreworld: Emergence and evolution of cooperative structures in a computational chemistry. *Physica D,* 1990, 42, 111–134.

[77] Reade, C. *Elements of Functional Programming.* Addison-Wesley, Reading, MA, 1989.

[78] Rétoré, C. Réseaux et Séquents Ordonnés, University of Paris VII, PhD Dissertation, 1993.

[79] Rokshar, D. S., P. W. Anderson and D. L. Stein. Self-organization in prebiological systems: Simulation of a model for the origin of genetic information. *J. Mol. Evol.,* 1986, 23, 110.

[80] Rosen, R. *Life Itself: A Comprehensive Inquiry into the Nature, Origin, and Fabrication of Life.* Columbia University Press, New York, 1991.

[81] Sangiorgi, D. Expressing Mobility in Process Algebras: First-Order

and Higher-Order Paradigms. University of Edinburgh, PhD Dissertation, 1992.

[82] Schönfinkel, M. Über die Bausteine der Mathematicischen Logik. *Math. Annalen,* 92 (1924), 305–316.

[83] Smith, B. C. *On the Origin of Objects.* Bradford Books, MIT Press, Cambridge, MA, 1996.

[84] Szathmáry, E. A classification of replicators and lambda-calculus models of biological organization. *Proc. R. Soc. Lond. B,* 1995, 260, 279–286.

[85] Thürk, M. Ein Modell zur Selbstorganisation von Automatenalgorithmen zum Studium molekularer Evolution. Universität Jena, Germany, PhD Dissertation, 1993.

[86] Troelstra, A. *Lectures on linear logic.* CSLI Lecture Notes 29, Center for the Study of Language and Information. Stanford, CA 1992.

[87] Varela, F. J. *Principles of Biological Autonomy.* North-Holland, New York, 1979.

[88] Varela, F. J., H. Maturana and R. Uribe. Autopoiesis: the organization of living systems, its characterization and a model. *BioSystems,* 5, 187–196, 1974.

[89] These materials are available in conjunction with the main text as IIASA Working Paper WP-96-27 or as Working Paper #41 from the Yale Center for Computational Ecology. They are also obtainable from the Web at `http://peaplant.biology.yale.edu:8001/alchemy.html` or `http://www.iiasa.ac.at/docs/IIASA_Publications.html`.

Chapter 5

SCIENTIFIC KNOWLEDGE FROM THE PERSPECTIVE OF QUANTUM COSMOLOGY

James B. Hartle

I. Introduction

The assignment of the organizers was to speak on the subject of "limits to scientific knowledge." This is not a topic on which I have been forced to reflect a great deal in the course of my efforts in astrophysics, but I shall try to offer a few thoughts on it from the perspective of cosmology. Like any assignment, the first task is to understand what it might mean. I shall say more about this later. But one thing is immediately clear: This is not simply an empirical question, but rather concerns the relationship between what we observe and our theories of what we observe. Limits therefore depend on theories and will vary from one scientific theory to another. The question of what are the *fundamental* limits to scientific knowledge must be examined in the most general theoretical context. In physics this is the subject of quantum cosmology – the quantum mechanics of the universe as a whole and everything inside it. The nature of scientific knowledge in this most comprehensive of theories is the subject of this chapter. I shall try to describe a little of what quantum cosmology is about, and address the question of limitations to scientific knowledge in this most general of contexts.

II. Three Kinds of Limits to Scientific Knowledge

In this section, three different kinds of limits to scientific knowledge are identified. No claim is made that these are the only kinds of limits. But these three have a general character that is inherent in the nature of the scientific enterprise. Subsequent sections will illustrate these general kinds of limits with examples from quantum cosmology.

A. Limits to What is Predicted

The task of science, as Bohr said, is "to extend the range of our experience and to reduce it to order" [1]. To reduce experience to order is to compress the length of a description of that experience. That compression is achieved when a computer program can be exhibited which, given certain input, produces a string describing some parts of our experience, and the length of that program together with its input are shorter than the length of the output description. Theory supplies the program. For instance, a detailed description of the observations of the positions of the planets over the last 100 years might make up a very long table. But Newton's equations of motion can be used to compress all that information into two much shorter strings: a string stating Newton's theory and another string giving the positions and velocities of the planets at one moment in time.

It is a logical possibility that *every* feature of our experience—the wave function of every quark, the velocity of every molecule, the position of every leaf, the character of each biological species, the action of every human—is just a very long output of a short computer program with *no* input. However, in the history of scientific inquiry there is no evidence that the universe is so regular. Even the most deterministic classical theories did not claim this. With Newtonian mechanics, Laplace proposed only to predict the future and retrodict the past *given* the present position and velocity of each particle in the universe. That list of initial data would be vastly longer than a few treatises on Newtonian mechanics. Existing theories predict a string describing our experience only given some other, shorter, string as input. Theories do not predict everything that is observed, but only certain regularities in our observations. Some things are predicted, some are not, and that limit to what is predicted is one kind of limit to scientific knowledge.

Scientific laws must have some degree of simplicity to be discoverable, comprehensible, and effectively applicable by human beings and other complex adaptive systems. If the complexity of the present universe is large, then this necessary simplicity of the laws implies that this kind of limit to scientific knowledge is inevitable. Not everything can be predicted but only those regularities that are summarized in the laws of science. In the following we shall describe what is predicted and what is not predicted in quantum cosmology.[1]

B. Limits to Implementation

To be tested, the predictions of an abstractly represented theory covering a broad class of phenomena must be implemented in particular circumstances. The theory must produce numbers, and that process involves computation. Even if the laws are precisely specified and even if the input to those laws is exactly stated, limitations of our ability to compute may limit our ability to predict. This is another kind of limit to scientific knowledge. The practical limitations of present computing machines are all too familiar. Computing the motion of every particle in a classical gas of 10^{22} particles in less than its real evolution time is well beyond the powers of contemporary computers. However, beyond the limitations of contemporary machinery, we may ask whether there are fundamental limitations on what can be computed that are inherent in the form of the laws themselves. The phenomenon of chaos is the source of one kind of limitation. The precision required of initial data to extrapolate a given time into the future increases exponentially with that time for a wide variety of classical systems. Another kind of limit arises in cosmology, where resources for computation, both in time and space, are limited. Further, as we shall see, there is some evidence that certain predictions of quantum cosmology may be non-computable numbers.

It is not difficult to display predictions that are computationally intractable but which are measurably inaccessible. Given initial conditions, classical theory predicts the orbits of every molecule of gas in a room. The explicit computation of this prediction at the operating speeds of current computers would take much longer than the age of the universe because of the large number of particles involved. Yet, for the same reason, neither the initial condition nor the predicted orbits are measurably accessible quantities. Merely exhibiting phenomena that are non-computable or computationally intractable is not much of a limit if the phenomena are impossible or extraordinarily difficult to measure. The most interesting limits concern phenomena that are easy to measure but difficult to compute.

---

[1]For a lucid discussion in popular language of the notion of complexity and of prediction in quantum cosmology, as well as a summary of some of the author's work with M. Gell-Mann, see [2].

C. Limits to Verification

The above discussion has assumed that we know the laws of physics. However, we arrive at those laws by a process of induction and experiment. Competing laws consistent with known regularities are winnowed by the process of checking their predictions with new observations. Are there fundamental limits to what we can test, and therefore fundamental limits to how well the theory can be known? Cosmology will provide examples.

D. False Limits

Beware of false limits that arise only from imprecise language or the comparison of a correct theory with an incorrect one. A classic example is provided by the uncertainty principle in quantum mechanics,

$$\Delta x \, \Delta p \geq \hbar/2. \tag{2.1}$$

This relation is sometimes described as a limit on our ability to predict (or "measure" or "know") both the position and momentum of a particle at one time to accuracies better than those restricted by the relation (2.1). However, the uncertainty principle is more accurately characterized as a limit on the use of classical language in a quantum mechanical situation.

There is no state of a quantum mechanical particle with a precisely defined position and momentum. That is the content of (2.1). The uncertainty principle, therefore, is not a limit to what observed properties of a quantum particle are predicted by the theory. Since there is no quantum state with precisely defined position and momentum, quantum theory predicts that we shall never observe both simultaneously. Thus, as far as the position and momentum of a particle are concerned, there is no disparity within quantum theory between what can be predicted and what is observed arising from (2.1), as there would be in the case of a genuine limit of the type discussed in Section A.

As mentioned earlier, limits to prediction are properties of the theories that specify what can be predicted. Of course, if we *compare* two theories one may predict different phenomena from the other. In classical physics there are states in which the position and momentum of a particle *are* simultaneously specified; in quantum theory there are not. But quantum theory is correct and classical theory incorrect for the domain of phenomena we have in mind. The uncertainty principle (2.1) may be viewed as a kind of limit to how far classical concepts and language can be applied in quantum theory. But, were we to strictly adhere to the language and concepts of quantum theory, it would be no limit at all.

III. Dynamical Laws and Initial Conditions

As mentioned above, fundamental limits to what is predicted, to how predictions can be implemented, and to how theory can be verified depend on what the fundamental theory is. This section sketches some essential features of fundamental physical theory today. Of course, we are on dangerous ground here. The most fundamental laws are often the furthest from definitive experimental test. Nevertheless, it is interesting to see what types of limits might exist in the kind of fundamental theoretical framework that is under active investigation by physicists today.

The most general framework for prediction is quantum cosmology—the quantum theory of the universe as a whole and everything that goes on inside it. In the following I shall describe a little of this theory.

Historically, physics for the most part has been concerned with finding dynamical laws—laws that compress the description of evolution over time to the description of an initial condition. Thus, these dynamical laws require boundary conditions to yield predictions. There are no particular laws governing these boundary conditions. They are specified by our observations of the part of the universe outside the subsystem whose dynamics is of interest. If we observe no incoming radiation in a room, we solve Maxwell's equations there with no-incoming-radiation boundary conditions. If we prepare an atom in a certain atomic state, we solve Schrödinger's equation with that initial condition, and so on.

But in cosmology we are confronted with a fundamentally different kind of problem. Whether classical or quantum, the dynamical laws governing the evolution of the universe require boundary conditions. But in cosmology there is no "rest of the universe" to pass their specification off to. The boundary conditions must be part of the laws of physics themselves; there is no other place to turn.

A current view, therefore, is that the most general laws of physics involve two elements:

- The laws of dynamics prescribing the evolution of matter and fields and consisting of a unified theory of the strong, electromagnetic, weak, and gravitational forces.
- A law specifying the initial boundary condition of the universe.

There are no predictions of any kind that do not depend on these two laws, even if only very weakly, or even when expressed through phenomenological approximations to these laws (like classical physics) that are appropriate in particular and limited circumstances with forms that may be only distantly related to those of the fundamental theory.

The search for a fundamental theory of the dynamics of matter has been seriously under way since the time of Newton. Classical mechanics, Newtonian gravity, electrodynamics, special relativity, general relativity, quantum mechanics, quantum electrodynamics, the theory of the electroweak interactions, quantum chromodynamics, grand unified theories, and superstring theory are but some of the important milestones in this search. By way of contrast, the search for a theory of the initial condition of the universe has been seriously under way for not much more than a decade. (See [3] for a review.) The reason for this difference can be traced to the scales on which the regularities summarized by these two laws emerge. The trajectory of a ball in the air, the flow of water in a pipe, or the motion of a planet in the solar system all exhibit the regularities implied by Newtonian mechanics. The regularities of the dynamical laws of atomic and particle physics can be exhibited in experiments carried out in laboratories or large accelerators. However, characteristic regularities implied by a theory of the initial condition of the universe emerge mostly on much larger, cosmological scales.

On any scale, the universe exhibits some regularities in space as distinct from regularities in time. Rocks on one part of the earth are related to rocks on another part. Similarly, there are relations between individual members of biological species and human history in different locations. These regularities have their origins in the common origin of rocks in the earth, the evolution of biological species, and the facts of human history. On cosmological scales the universe is more regular in space than it is on smaller scales. The progress of observation in astronomy in recent decades has given us an increasingly detailed picture of the universe on ever larger scales of space and time. The remarkable inference from these observations is that the universe becomes increasingly simple as we move to larger scales in space and more distant times in the past.

Galaxies are not very complicated objects but still exhibit a variety of types and considerable individuality. On the larger scale of a tenth of the radius of the universe, the galaxies are no longer individual objects but display considerable structure in their distribution. Pictures of the distribution of galaxies on the sky, which probe out to greater distances show less structure. On the largest scales, the distribution of the cosmic background radiation temperature, which is as close as we can come to a picture of the whole universe 300,000 years after the big bang, reveals almost no structure at all. (See Figure 1.) The deviations in this temperature from exact smoothness (exact isotropy) are measured in tens of *millionths* of a degree. But those deviations are important! They are the origin of all the complexity in the universe we see today. As the universe evolves, these fluctuations grow, collapse, and fragment through gravitational attraction to become the galaxies, stars, and planets that characterize the universe today. Initially very close to equilibrium, the matter in the universe is thereby driven further from equilibrium. That is the disequilibrium necessary for chemistry, geology, life, biology, and human history.

The evidence of the observations then is that the universe was a simpler place earlier than it is now—more homogeneous, more isotropic, with matter much closer to thermal equilibrium. The aim of quantum cosmology is a quantum theory of this simple initial condition.

IV. Classical and Quantum Initial Conditions

It is an inescapable inference from the physics of the last sixty years that we live in a quantum mechanical universe—a world in which the basic laws of physics conform to that general framework for prediction we call quantum mechanics. We perhaps have little evidence for peculiarly quantum mechanical phenomena on large and even familiar scales, but there is no evidence that the phenomena that we do see cannot be described in quantum mechanical terms and explained by quantum mechanical laws. This is the first reason that the search for a theory of the initial condition is carried out in the framework of *quantum* cosmology. There is, however, another reason: quantum indeterminacy is probably necessary for a comprehensible fundamental, scientific theory of the initial condition.

To explain this necessity and also to understand a bit of the machinery of quantum cosmology, consider a model universe. Suppose the universe consists of a box the size of the visible universe containing a large number N of particles interacting by fixed potentials. To simulate the expansion of the universe we could let the box expand. That's actually not a bad model for what goes on in more recent epochs of the universe.

Classically a history of this model universe is a curve in a $6N$-dimensional phase space of the positions and momenta of all the particles in the box. Classical evolution is deterministic—if the point in phase

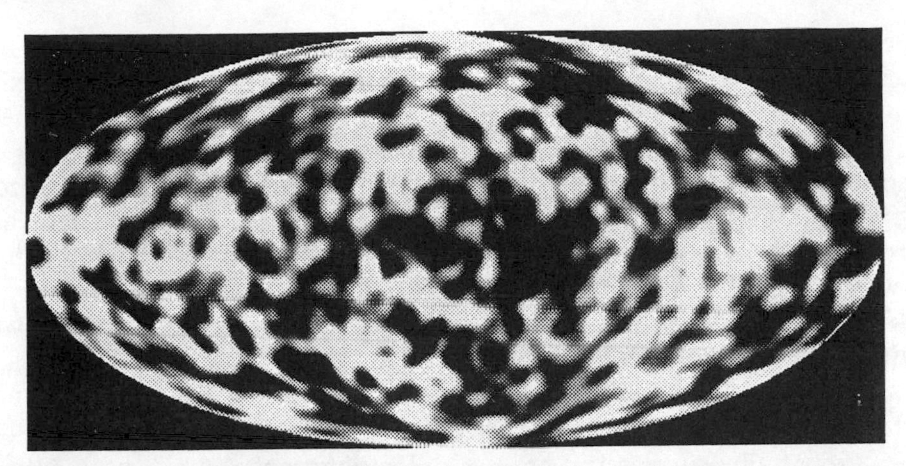

Figure 1. A sky map of temperature fluctuations in the cosmic background radiation. This figure may be thought of as a picture of the universe approximately 300,000 years after the big bang. The hot mixture of matter and radiation that exists immediately after the big bang cools as the universe expands. About 300,000 years later the universe has cooled enough that matter and radiation no longer significantly interact. Photons from that time have been traveling freely towards us ever since. Their characteristic temperature now is only 2.7 degrees above absolute zero, yet they can be detected at microwave wavelengths by sensitive instruments. The figure above shows a sky map of the temperature of that radiation based on data taken with the COBE satellite. The dark spots are where the sky is cooler than the mean temperature and the white areas are where it is hotter. The differences in temperature between the darkest black and the whitest white is only a few hundred *micro* degrees Kelvin. The universe is thus essentially featureless at 300,000 years after the big bang except for these tiny fluctuations. These small fluctuations, however, are the origin of all the complexity in the universe that we see today. [Greyscale adaptation by J. Gundersen of the results of C. Bennett, et. al. *Ap. J. 436* (1994), 423]

space specifying the system's configuration is known at one time, the location at all other times is determined by the equations of motion. A classical theory of the initial condition of the model universe thus might specify the initial point in phase space at $t = 0$. However, such a theory would necessarily be hopelessly complex because it would have to encode all the complexity we see today. Its description would be too long to be comprehensible.

A *statistical* classical initial condition could be simpler. Such an initial condition would only give a *probability* for the initial point in phase space and therefore only a probability for the subsequent evolution. Present predictions of the future would then be probabilistic. For example, observers at any time in the history of the universe can only see galaxies within a distance close enough that their light could have reached them in the time since the big bang. This cosmological horizon expands as the universe ages. One new galaxy comes over this cosmological horizon approximately every 10 minutes. A statistical initial condition might not predict with near certainty, say, the specific locations of the individual new galaxies, but rather would forecast their statistical distribution on the sky. Similarly, with a classical initial condition in which matter was initially in thermal equilibrium, one might predict the overall intensity of the background radiation on the sky, but not the location of any particular fluctuation in its intensity.

Probabilities in classical physics reflect ignorance. A classical statistical law of the initial condition would mean that we have some information about how the universe started out, but not all. However, we learn from observation. With every observation we could refine our theory of the initial condition which would therefore become increasingly complex, reflecting the complexity of the present, and thus become increasingly less comprehensible.

Quantum mechanics is inherently indeterministic and probabilities are basic. The most complete specification of the initial state of our model box of particles would be a wave function on the configuration space of all their positions—a wave function of the universe, $\Psi(\vec{x}_1, \cdots, \vec{x}_N)$, for this model.

Unlike classical physics, subsequent observation will not improve this initial condition, although the results of observation can be used to improve future predictions. Thus, it is natural in quantum mechanics to have a simple, comprehensible law of the initial condition which is consistent with the complexity observed today.[2]

V. What is Predicted in Quantum Cosmology?

My colleague, Murray Gell-Mann, once asked me, "If you know the wave function of the universe, why aren't you rich?" The answer is that very little is predicted with certainty by such a quantum initial condition of the universe and what is predicted is certainly not of much use in generating wealth. What might be predicted by an initial condition for cosmology is the subject of this section.

---

[2]For an early statement of this, see [4].

Quantum mechanics predicts probabilities for sets of alternatives. In our model universe in a box, for example, it might predict the probabilities for alternative ranges of the position of a particle at a particular time, or the probabilities for alternative distributions of energy density in the box, and many other sets of alternatives. These are the probabilities for alternatives that are *single* events in a *single* closed system — the universe as a whole.

What do such probabilities of single events mean? Some may find it helpful to think of these probabilities as predictions of relative frequencies in an imaginary infinite ensemble of universes. But they are not frequencies in any accessible sense. Rather, to understand what the probabilities of single events mean, it is best to understand how they are used. Probabilities of single events can be useful guides to behavior even when they are distributed over a set of alternatives so that none is very close to 0 or 1. Examples are the probability that it will rain today or the probability of a successful marriage. However, because the probabilities are distributed, the event which occurs — rain or no rain, divorce or death before parting — does not test the theory that produced the probabilities. Tests of the theory occur when the probabilities are *near certain,* by which I mean sufficiently close to 0 or 1 that the theory would be falsified if an event with probability sufficiently close to 0 occurred, or an event with a probability sufficiently close to 1 did not occur.[3]

Various strategies can be used to identify sets of alternatives for which probabilities are near 0 or 1. The most familiar is to study the frequencies of outcomes of repeated observations in an ensemble of a large number of identical situations. Such frequencies would be predicted with certainty in an infinite ensemble. However, since there are no genuinely infinite ensembles in the world, we are necessarily concerned with the probability for the deviations of the frequency in a finite ensemble from the expected behavior of an infinite one. Those are probabilities for single properties (the deviations) of a single system (the whole ensemble) that become closer and closer to 0 or 1 as the ensemble is made larger.

Another strategy for identifying alternatives with probabilities near 0 and 1 is to consider probabilities conditioned on other information besides that given in the theory of dynamics and the initial condition of

---

[3]*How* close to 0 or 1 probabilities must be for near certain predictions depends on the circumstances in which they are used as I have discussed elsewhere [5].

the universe. Present theories of the initial condition do not predict the observed orbit of Mars about the sun with any significant probability. But they do predict that the *conditional* probability for the observed orbit is near 1 *given* a few previous observations of Mars' position. Such conditional probabilities are what are used in the rest of the sciences when they are viewed from the perspective of quantum cosmology as we shall discuss in more detail in the subsequent sections.

In the following discussion it will be helpful to use just a little of the mathematics of quantum mechanics to discuss quantum cosmology.[4] For simplicity and definiteness, let us continue to discuss the model universe of N particles in a box. The quantum initial state of this model universe is represented by a state vector $|\Psi\rangle$ in a Hilbert space, or equivalently by a wave function of the coördinates of all the particles in the box:

$$\Psi(\vec{x}_1, \cdots, \vec{x}_N). \tag{5.1}$$

General alternatives at a moment of time whose probabilities we might want to consider can always be reduced to a set of "yes-no" alternatives. For instance, questions about the position of a particle can be reduced to questions of the form: "Is the particle in this region—yes or no?", "Is the particle in that region—yes or no?", and so on. A set of "yes-no" alternatives at one moment of time, say $t = 0$, is represented by a set of orthogonal projection operators $\{P_\alpha\}, \alpha = 1, 2, \cdots$—one projection operator for each alternative. (A projection operator is one whose square is equal to itself.) The projection operators satisfy

$$\sum\nolimits_\alpha P_\alpha = I \quad \text{and} \quad P_\alpha P_\beta = 0, \quad \alpha \neq \beta, \tag{5.2}$$

showing mathematically that they represent an exhaustive set of exclusive alternatives. The same set of alternatives at a later time $t > 0$ is represented by a set of (Heisenberg picture) projection operators $\{P_\alpha(t)\}$. The time dependence of each $P_\alpha(t)$ is given by

$$P_\alpha(t) = e^{iHt} P_\alpha e^{-iHt}, \tag{5.3}$$

where H is the Hamiltonian encapsulating the basic dynamical theory. The probability predicted for alternative α at time t is

$$p(\alpha) = \left\| P_\alpha(t) |\Psi\rangle \right\|^2, \tag{5.4}$$

---

[4]For more details at an elementary level see [6], and in greater depth see [7].

where $\|\chi\rangle\|$ means the length of the Hilbert space vector $|\chi\rangle$. For this model, the Hamiltonian H specifies the first of the two elements of a basic physical theory described in Section III—the fundamental theory of dynamics. The state vector $|\Psi\rangle$, or equivalently the wave function $\Psi(\vec{x}_1, \cdots \vec{x}_n)$, specifies the second element—the initial condition.

Probabilities for alternatives at a moment in time are not the most general predictions of quantum mechanics. More generally, one can ask for the probabilities of sequences of sets of alternatives at a series of different times $t_1 < t_2 < \cdots < t_n$ making up a set of alternative *histories* for the universe. Each history corresponds to a particular sequence of alternatives $(\alpha_1, \cdots, \alpha_n)$, and is represented by an operator that is the chain of projections corresponding to the sequence of alternatives

$$C_\alpha = P^n_{\alpha_n}(t_n) \times \cdots \times P^1_{\alpha_1}(t_1). \tag{5.5}$$

Here, the index α is shorthand for the whole sequence $(\alpha_n, \cdots, \alpha_1)$, while the superscripts on the P's indicate that different sets of alternatives can be considered at different times. When the operator C_α is applied to the initial state vector $|\Psi\rangle$, one obtains the *branch state vector* $C_\alpha|\Psi\rangle$ corresponding to the history α. The probability of the history α is the length of the history's branch state vector:

$$p(\alpha) = \|C_\alpha|\Psi\rangle\|^2. \tag{5.6}$$

Probabilities of histories are essential for predicting such everyday things as the orbit of the moon, which is a sequence of positions at a series of times.

We can now begin to analyze the question of what is and is not predicted in quantum cosmology. The most characteristically quantum mechanical limitation on what can be predicted is that not every set of alternative histories that may be described can be assigned probabilities by the theory. This is because of quantum mechanical interference, which is very clearly exemplified in the two-slit thought experiment illustrated in Figure 2. Electrons proceed from an electron gun through a barrier with two slits on their way to detection at a screen. Passing through slit A or slit B defines two alternative histories for the electrons arriving at a fixed point y on the screen.

In the usual story, if we have not measured which slit an electron passed through, then it would be inconsistent to predict probabilities for these alternative histories, because the probability to arrive at y

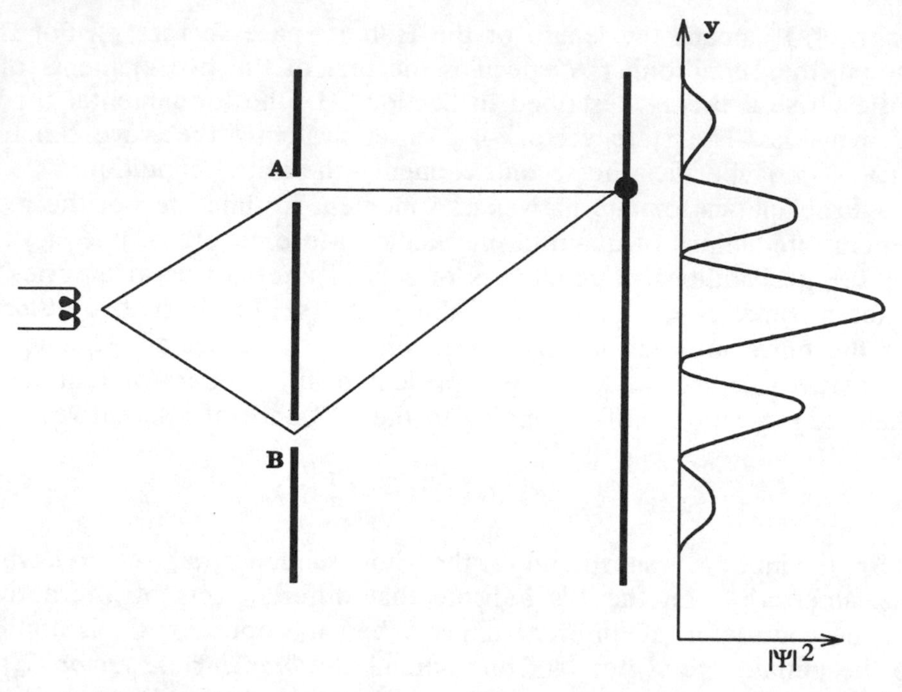

Figure 2. The two-slit experiment. An electron gun at left emits an electron traveling towards detection at a screen at right, its progress in space recapitulating its evolution in time. In between there is a barrier with two slits. Two possible histories of an electron arriving at a particular point on the screen are defined by whether it went through slit A or slit B. In quantum mechanics, probabilities cannot be consistently assigned to this set of two alternative histories because of quantum mechanical interference between them. However, if the electron interacts with apparatus that measures which of the slits it passed through, then interference is destroyed, the alternative histories decohere, and probabilities can be assigned to the alternative histories.

would not be the sum of the probability to pass through A to y and the probability to pass through B to y:

$$p(y) \neq p_A(y) + p_B(y). \tag{5.7}$$

The reason is that in quantum mechanics probabilities are the squares of amplitudes and

$$|\psi_A(y) + \psi_B(y)|^2 \neq |\psi_A(y)|^2 + |\psi_B(y)|^2. \tag{5.8}$$

It is not that we are ignorant of which slit the electron passes through, so that the probabilities are 50–50. Rather, it is inconsistent to discuss probabilities at all.

Therefore, quantum mechanics, in any of its various levels of formulation, contains a rule specifying which sets of alternative histories may be assigned probabilities and which may not. In the most general context of the quantum mechanics of the universe, that rule is as follows [8–10]: Probabilities may be consistently assigned to just those sets of histories for which there is vanishing interference between the individual members of the set as a consequence of the universe's initial state $|\Psi\rangle$. Such sets of histories are said to *decohere.* The condition for a decoherent set of histories is that the branches of the initial state $C_\alpha|\Psi\rangle$ corresponding to individual histories be mutually orthogonal:

$$\langle\Psi|C_\alpha^\dagger \cdot C_\beta|\Psi\rangle \approx 0, \quad \alpha \neq \beta. \tag{5.9}$$

As a result, the most general probability sum rules are satisfied. Consistency limits the predictions of quantum theory to the probabilities of *decoherent* sets of alternative histories.

As an example of how the decoherence of a set of histories comes about, suppose we have a single millimeter-sized dust grain in a quantum state that is a superposition of two positions about a millimeter apart located deep in intergalactic space. Consider alternative histories of the position of this particle at a sequence of a few times. (The P's in (5.5) would then be projections onto ranges of this position.) Were the particle isolated, this situation would be analogous to the two-slit experiment, and histories of differing positions would not decohere. However, even deep in space this particle is not isolated. The all-pervasive light from the big bang illuminates the particle, and about 10^{11} cosmic background photons scatter from it every second. Through these interactions, this seemingly isolated dust grain becomes correlated with radiation in a part of the universe whose size is growing at the speed of light. The two states with different positions become correlated with two different, nearly orthogonal, states of the radiation after a time of about a nanosecond. By this means, a branch of the initial state in which the grain is initially at one position becomes orthogonal to a branch in which the grain is a millimeter away. Decoherence of alternative histories of position has been achieved, since the relative phase between states of different position has been dissipated by feeble interactions with the background radiation. Mechanisms such this are widespread in the universe, and typical of those effecting the decoherence of histories of the kinds of classical variables we like to follow (see, e.g., [11,12]).

In the above example, decoherence of alternative histories of the position of the dust grain is achieved at the cost of ignoring the photons that are effecting the decoherence. That is an example of *coarse-graining.* Were we to consider a set of alternative histories of states of the cosmic background radiation, as well as the position of the grain, we would be, in effect, following all possible phase information. Such a set of alternative histories would generally not decohere. Except for trivial cases, sets of histories must describe *coarse-grained* alternatives in order for probabilities to be predicted at all. This necessary imprecision is a genuine limit to what can be predicted in quantum cosmology, in contrast to the limits of the kind associated with the uncertainty principle which are merely limits to the applicability of classical modes of description.[5] (For more details, see [7].)

We thus have the picture of a vast class of all possible sets of alternative histories and a smaller subclass of decoherent sets of histories for which quantum theory predicts probabilities. For almost none of these decoherent sets is there a history predicted with certainty on the basis of the initial state alone. If one history has probability 1, then all alternatives to it must have probability 0. Suppose we have such a set, and let α_c be the label of the certain history, then from (5.6),

$$\|C_\alpha|\Psi\rangle\|^2 = 0, \quad \alpha \neq \alpha_c, \tag{5.10}$$

which implies

$$C_\alpha|\Psi\rangle = 0, \quad \alpha \neq \alpha_c. \tag{5.11a}$$

Then, since $\Sigma_\alpha C_\alpha = I$ as a consequence of (5.2), we also have

$$C_\alpha|\Psi\rangle = |\Psi\rangle, \quad \alpha = \alpha_c. \tag{5.11b}$$

---

[5]Decoherence also implies another kind of limit to classical predictability which should be mentioned, although we cannot discuss it in any depth here. As described, realistic mechanisms of decoherence involve the dispersal of phase information concerning a subsystem into an environment that interacts feebly with it. Those interactions produce noise that limits the classical predictability of the subsystem. Thus, for classical predictability appropriate and sufficient coarse-graining is needed for the decoherence necessary to predict probabilities at all. But further coarse-graining is needed for the subsystem to have sufficient inertia to resist the noise that those mechanisms of decoherence produce and thereby become classically predictable. (For an introductory discussion see [11].)

Decoherence, Eq. (5.9), is then automatic for such sets of histories of which one is certain.

Equation (5.11b) shows that operators of histories that are predicted with certainty act as projection operators on the initial state. An alternative predicted with probability 1 is thus mathematically equivalent to the alternative corresponding to the question, "Is the universe in state $|\Psi\rangle$?" These are very special questions. Out of the class of sets of decoherent histories almost none correspond to sets in which one history is a certain prediction of the initial condition and the theory of dynamics alone.

In quantum cosmology we might hope that some gross features of the universe would be among those that are predicted with near certainty from the initial condition and dynamics alone. These include features such as the approximate homogeneity and isotropy of the universe on scales above several hundred megaparsecs,[6] its vast age after the big bang when measured on elementary particle time scales, and certain features of the spectrum of density fluctuations that grew to produce the galaxies. On more familiar scales, we may hope that the laws of the initial condition and dynamics would predict the homogeneity of the thermodynamic arrow of time and the wide range of scale and epoch on which the regularities of classical physics are exhibited. There has even been speculation that phenomena on very small scales, such as the dimensionality of spacetime or certain effective interactions of the elementary particles at accessible energy scales, may be near certain predictions of the initial condition and dynamics. But there is little reason to suspect that simple theories of the initial condition and fundamental dynamics will predict anything about the behavior of the New York Stock Exchange with near certainty, and a great many other interesting phenomena as well. That is why you can't get rich knowing the wave function of the universe!

The situation is very different if information beyond laws of dynamics and the initial condition is supplied and probabilities conditioned on that information are considered. There are many sets of *conditional* probabilities in which one member of the set is near certain. These conditional probabilities are the basis of prediction in all the other sciences when viewed from the perspective of quantum cosmology as will be described in the next section.

---

[6]A megaparsec (Mpc) is a convenient unit for cosmology. One megaparsec is 3.3 million light years, which equals about 3.1×10^{24}cm. The size of the universe visible today is of order several thousand megaparsecs.

I have described various limitations on what can be predicted in quantum cosmology. Yet there is a sense in which we, as information gathering and utilizing systems, make use of only a small part of the possible predictions of quantum cosmology. That is because of our almost exclusive focus on alternatives defined in terms of the variables of classical physics—averages over suitable volumes of densities of energy and momentum, densities of nuclear and chemical species, average field strengths, and so forth. Such classical quantities are represented by quantum operators called *quasiclassical operators*. (They are termed "*quasi*classical" because they do not behave classically in all circumstances.) Certainly our immediate experience can be described in terms of quasiclassical variables even when—as in the clicks of a Geiger counter—these variables do not obey deterministic classical laws.

Even in our theorizing about regions of space or epochs in time that are very distant from us, we often focus on histories of alternatives of quasiclassical operators. Only in the microscopic arena do we consider non-quasiclassical alternatives, such as electron spin and coherent superpositions of position. Even then, we typically consider such alternatives only when they are tightly correlated with a quasiclassical variable as in a measurement situation.

However, quantum field theory exhibits many more kinds of variables than the small set of quasiclassical ones. Decohering sets of histories can be constructed from alternative values of non-quasiclassical operators as well as from quasiclassical ones. Indeed, the quasiclassical sets of histories are but a small subset of the whole class of decohering histories. Quantum theory does not confer privilege on any one set of decohering histories over another. Probabilities are predicted for all such sets of alternatives. Histories of non-quasiclassical alternatives are not beyond reach, however.

Suppose we were to make measurements of peculiarly quantum mechanical variables involving large numbers of particles in regions of macroscopic dimensions. The histories that would be relevant for the explanation of the outcomes of these measurements would not be histories of quasiclassical variables in these regions, but rather histories of the non-quasiclassical alternatives that were measured. The reason for our preference for quasiclassical sets of alternative histories, like all other questions concerning ourselves as particular physical systems, probably lies in our evolutionary history—not in the framework of quantum theory itself.

VI. Differences Between the Sciences

Using the conditional probabilities of quantum cosmology, a particular orbit of the earth about the sun could be predicted with near certainty given a few previous positions of the earth and a description in terms of the fundamental fields of the earth and solar system which are the language of quantum cosmology. The probabilities for the outcome of chemical reactions become near certain predictions of quantum cosmology, given a description in terms of fundamental fields of the molecules involved and the conditions under which they interact. The probabilities for the behavior of sea turtles in particular environments could, in principle, become predictions of quantum cosmology, given a description of sea turtles and their environments in the language of quantum cosmology. Even the probabilities for the different behaviors of human beings—both individually and collectively—could in principle be predicted given a sufficiently accurate description of the individuals, their history, their environment, and their possible modes of behavior. In this way, *every* prediction in science could be viewed in terms of a conditional probability in quantum cosmology. Why then do we have separate sciences of astronomy, chemistry, biology, psychology, and so on? The answer, of course, is that it is neither especially interesting nor practical to reduce the predictions of these sciences to a computation in quantum cosmology.

One measure of the difference between the sciences is how sensitive the regularities they study are to the forms of the initial condition of the universe and the fundamental theory of dynamics. The phenomena studied in chemistry, fluid mechanics, geology, biology, psychology, and human history, depend only very little on the particular form of the initial condition. All of these sciences, especially chemistry, depend on the form of the theory of dynamics in some approximation. But as we move through the list we are moving in the direction of the study of the regularities of *increasingly specific subsystems of the universe.* Specific subsystems can exhibit more regularities than are implied generally by the laws of dynamics and the initial condition. The explanation of these regularities lies in the origin and evolution of the specific subsystems in question. Naturally, these regularities are more sensitive to this specific history than they are to the form of the initial condition and dynamics. That is especially clear in a science like biology. Of course, living systems conform to the laws of physics and chemistry, but their detailed form and behavior depend much more on the frozen accidents of several billion years of evolutionary history on a particular planet moving around a particular star than they do on the details of superstring theory or the "no-boundary" initial condition of the universe.[7] Conversely the phenomena studied by these sciences do not help much in discriminating among different theories of the initial condition and dynamics. It is for such reasons that it is not of pressing interest—either for other areas of science or for quantum cosmology itself—to express the predictions of such phenomena as quantum cosmological probabilities, even though it is in principle possible to do so.

Even if we wanted to carry out a calculation of the conditional probabilities in quantum cosmology necessary for prediction in the other sciences, an examination of what it would take yields three measures that distinguish the other sciences from quantum cosmology and from one another. To yield a conditional probability the theory requires:

1. A description of the coarse-grained alternatives whose probabilities are to be predicted in terms of fundamental quantum fields.
2. A description of the circumstances under which the probabilities are conditioned in terms of fundamental quantum fields.
3. A computation of the conditional probabilities.

Table 1 shows some simplistic guesses of the lengths of these three parameters for typical problems in the various sciences. We can discuss a few of these.

By classical physics, I mean simply Newton's laws of mechanics and gravity, the laws of continuum mechanics, Maxwell's electrodynamics, the laws of thermodynamics, etc.—in short, the basic laws of physics as they were formulated in the 19th century. (I do not mean some specific application of these laws, as to the breaking of ocean waves.) Classical physics might almost be counted as a science separate from physics, for the laws of classical physics do not hold universally, but only for certain kinds of subsystems in particular circumstances. However, the table shows the reason these laws are usually considered part of the science of physics. There is just a short list of quasiclassical variables (volume averages of fields, densities of energy, momentum, chemical composition, etc.) whose ranges of values define the coarse-grained alternatives of classical physics [7]. It is a somewhat longer business to spell out in quantum mechanical terms the circumstances in which classical physics applies.

[7]See [2] for more discussion and examples from this point of view.

Table 1. Some Differences Between the Sciences

	Length of coarse-grained		
	description of alternatives	description of conditions	Computation of conditional probabilities
classical physics	very short	short	very short
astronomy	short	short	short–long
fluid mechanics	short–long	short	short–long
chemistry	short–long	short–long	long–very long
geology	long	long	long
biology	long–very long	long–very long	long–very long
psychology	very long	very long	very, very long(?)

But the derivation of the laws of classical physics can be as short as a journal paper.[8]

As we move down the table to astronomy, we encounter more specific classes of physical systems — stars, clusters, and galaxies. However, the difficulty of obtaining data on such distant objects prevents us from learning much individual detail. The length of the coarse-grained descriptions of both conditions and alternatives are typically short. The computations utilizing the equations of classical physics, however, range from very short dimensional estimates to long simulations of supernovae explosions.

In fluid mechanics we encounter a wide variety of particular phenomena arising from differential equations of classical physics. One has only to mention laminar flow, turbulence, cavitation, percolation, convection, solitons, shock waves, detonation, superfluidity, clouds, dynamos, internal waves, ocean waves, the weather, and climate, to recall something of the richness of phenomena studied in this subject. The coarse-grainings describing the different behaviors of fluids can sometimes be long, although the description of the conditions is usually shorter. Many of these phenomena can be simulated on today's computers by solving the differential equations of classical physics. These calculations could be considered calculations in quantum cosmology were we to append a standard description of the alternatives and conditions to them, together with computations that justify the use of these approximate equations

[8]For a one-journal-paper derivation from the quantum cosmological point of view see, e.g., [7].

from the basic theory of quantum fields and the initial condition under these conditions.

The description of the molecules of interest in chemistry can vary from short—as in typical chemical formulae—to long—as in the base sequence in human DNA. There is a similar range of conditions for chemical reactions ranging from a few reagents in a test tube to the interiors of cells. Quantum chemists *can* compute certain chemical properties, such as the those of chemical bonds, directly from the equations of an effective low-energy theory of the elementary particles. But these computations can only be described as long.

In geology we have a science concerned with a very specific system—the earth—observed in considerable detail. A lengthy string is needed to describe the alternative configurations and composition of the material on the surface in the detail that we know it. A long history would have to be described to set the conditions for calculating the probabilities and calculations of these probabilities. And even assuming the laws of classical physics, such a string would be very long.

The reader probably needs little convincing that the description of the behavior of a complex biological organism plus its evolutionary history and its present environment in the language of quantum field theory would be a long business indeed! We should not pretend that we are anywhere close to being able to give such a description or to being able to carry out the relevant computations of conditional probabilities in quantum cosmology. Psychology and human history are yet more difficult. We may have a rough idea of how to describe the action of a bird's beak in the language of quantum field theory, but very little idea of the coarse grainings that describe an individual's thoughts and emotions or the vicissitudes of empires.

Dear reader: Please do not write the author concerning the inadequacies of the above discussion. He is aware that the boundaries between the sciences are not precisely defined, and that there is wide variation in these three parameters within each one. In astronomy, for example, the description of our nearest star—the sun—can be just as complex as that of any phenomena in fluid mechanics (and indeed is a part of fluid mechanics). It may be possible to simulate the smallest self-reproducing biological units by conceivable computers [13]. There may be universal principles of mind that derive rather directly from the basics of physics [14]. The important point is that, at a fundamental level, every prediction in science may be viewed as the prediction of a conditional probability for alternatives in quantum cosmology, and that the probabilities relevant to different sciences may be distinguished, in part, by their sensitivity to the theories of the initial condition and dynamics, by the length of the description of the alternatives, by the length of the description of the conditions, and by the length of the computation needed to produce them.

VII. Limits to Implementation

The preceding section discussed some limitations of practice in our effort to implement the predictions of quantum cosmology for interesting specific subsystems in the universe. These limits were of the general character described in Section IIB. Are there more fundamental and general limits arising from computational intractability? Quantum cosmology provides some examples.

There are both physical and mathematical reasons for computational intractability. Landauer [15] has raised the issue of whether there are predictions whose computation would require more resources in space, time, and matter than are available in the universe. Quantum cosmology may also present an example of what might be regarded as an extreme example of mathematical computational intractability. There is some evidence that the wave function of the universe might be non-computable in the technical mathematical sense.

One idea for a theory of the wave function of the universe is the "no-boundary" proposal [16]. To understand a little of this idea, assume, for simplicity, that the universe is spatially closed and that gravity is the only quantum field. A cosmological wave function is then a function of the possible geometries of three-dimensional space. The "no-boundary" idea is that the value of the wave function of the universe Ψ at one particular spatial geometry is a sum over all locally Euclidean four-dimensional geometries having this three-dimensional space as a boundary and *no other boundaries.* Each four-dimensional geometry $\mathcal{G}$ in the sum is weighted by $\exp(-I[\mathcal{G}])$, where $I[\mathcal{G}]$ is the classical action for the geometry. Mathematically, a geometry is a specification of a notion of distance (a metric) on a space such that any small region can be smoothly mapped to a region of flat Euclidean space (a manifold). A sum over geometries would therefore naturally include a sum over manifolds as well as a sum over metrics. By suppressing two of the four dimensions, we can give the crude pictorial representation of this double sum as shown in Figure 3.

The mathematics of quantum gravity has not been developed to the point that we have a precise mathematical formulation of what the

Figure 3. The wave function of the universe as sum over manifolds and metrics. This figure uses two-dimensional analogs to illustrate some of the ideas that enter into the construction of the "no-boundary" wave function of the universe. That wave function is a function of three-dimensional spatial geometries, one of which is represented here in two fewer dimensions by the heavy circular curve. For that given three-geometry, the "no-boundary" wave function is a sum over Euclidean four-geometries that have it as one boundary and *no other boundary.* This sum can be divided into a sum over four-manifolds and a sum over different four-metrics on those manifolds. The two-dimensional analog of this sum is shown above. The surfaces in each column represent different metrics on the *same* manifold. The manifolds in each column are the same because the surfaces can be smoothly deformed into one another by changing their shape. The metrics are different from one surface to another in a given column because the distance between two points is generally different from one shape to another. For example, the overall surface area may differ from one shape to another. The two-dimensional surfaces in different columns are different manifolds because they have different numbers of handles, and surfaces with different number of handles cannot be smoothly deformed into one another. A sum over manifolds is thus analogous to the sum over columns. A sum over metrics is analogous to the sum over different surfaces in each column.

relation schematically represented in Figure 3 might mean. One idea for making it precise is to approximate each term in the sum by a manifold constructed of flat four-simplices—the four-dimensional analogs of triangles in two-dimensions and tetrahedra in three-dimensions. The two-dimensional analog of such a simplicial manifold is a surface made up of triangles, like the geodesic dome illustrated in Figure 4. To calculate in four dimensions the sum pictured crudely in Figure 3, one proceeds as follows: Choose a large number of four-simplices N. Find all possible manifolds that can be made by joining these four-simplices together. Choose *one* such assembly to represent each manifold in the sum. To approximate the sum over metrics, integrate $\exp(-I[\mathcal{G}])$ over the edge lengths of the simplices that are compatible with the triangle and similar inequalities. Sum the result over all manifolds. Take the limit as $N \to \infty$. That is one possible way the sum over geometries in the "no boundary" proposal for the wave function of the universe might be implemented.[9]

A computer program to carry out this task would first have to try all possible ways of assembling N four-simplices together and reject those that do not give a manifold. This is already a formidable mathematical problem, and it has only recently been shown that an algorithm exists to carry out this computation for four-dimensional manifolds [18–19]. The next step would be for the computer to take this list of four-manifolds and eliminate duplications. However, it is known that the issue of whether two simplicial four-manifolds are identical is undecidable.[10] More precisely, there does not exist a computer program which, *for any* N, can compare two input assemblies of N four-simplices making up manifolds, and halt after having printed out "yes" if the manifolds are identical and "no" if they are not.

This suggests that the wave function of the universe defined by a sum over geometries that includes a sum over manifolds is a non-computable number.[11] However, appearances can be deceptive. Whether a number is non-computable or not is a property of the number and not of the way the number is represented. Merely exhibiting one non-computable representation, like the series in Figure 3, does not establish that there is not some other representation in which it *is* computable. Demonstrating non-computability in such cases is likely to be a difficult mathematical problem.

---

[9]For more details and references to the earlier literature see, e.g., [17].

[10]For a review see [20].

[11]We are specifically assuming the Turing model of computation.

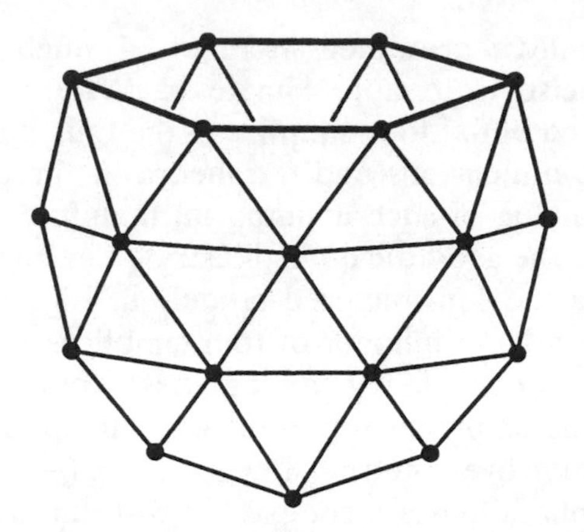

Figure 4. A smooth two-dimensional surface may be approximated by an assembly of flat triangles like a geodesic dome. In an analogous way, a curved four-dimensional geometry may be approximated by an assembly of four-simplices—the four-dimensional analog of two-dimensional triangles or three-dimensional tetrahedra. The four-simplices must be assembled so as to make a manifold—a space such that any small region can be smoothly mapped to a region of Euclidean space. Suppose we are given N four-simplices. Assembling them in different ways can give different manifolds (the different vertical columns in Figure 3). Different assignments of lengths to the edges give different sizes and shapes on a given manifold, that is, a different metric (the different shapes within each column in Figure 3). A sum over geometries, which is a sum over manifolds and metrics, may therefore be approximated by choosing one assembly to represent each manifold in the sum, integrating over its possible edge-lengths, and summing over all manifolds. The resulting sum may be a non-computable number because there is no algorithm for deciding when two simplicial four-manifolds are identical.

This suggested non-computability of sums over topologies has been taken as motivation for modifying the theory of the initial condition so that it is clearly computable [21–22]. But suppose that the wave function of the universe were non-computable. What would be the implications for science? Bob Geroch and I analyzed the implications of non-computability for physics in 1986 [23]. Our conclusion was that the prediction of non-computable numbers would not be a disaster for physics, because at any one time one needs theoretical predictions only to an accuracy consistent with experimental possibilities. Suppose, for example, it was sufficient for comparison with present observations to know the wave function of the universe to an accuracy of 10%. Suppose further it could be shown that to achieve this accuracy only simplicial manifolds with less than 100 four-simplices need be included in the series defining the wave function. The theorem concerning the non-existence of an algorithm for deciding the identity of simplicial four-manifolds refers to an algorithm that would work for *any* N. It does not rule out establishing the identity of two four-manifolds with less than 100 four-simplices. Indeed, being a problem that involves a finite number of specific cases, one imagines it could be solved with sufficient work on those cases. What the theorem ensures is that, if observations improve, and the wave function is later needed to an accuracy of 1%, requiring manifolds with a larger number of four-simplices (say 10,000), a new intellectual effort will be required to compute it. The algorithms that worked for manifolds assembled from less than 100 four-simplices are unlikely to work for manifolds assembled from less than 10,000 four-simplices.

Thus, the prediction of non-computable numbers would not mean the end of comparison between theory and observation. It would mean that the process of computing the predictions could be as conceptually challenging a problem as posing the theory itself.

VIII. Limits to Verification

Quantum mechanics predicts the probabilities of alternative histories of the universe. We cannot interpret these probabilities as predictions of frequencies that are accessible to test, for we have access to but a single universe and but a single history of it. Our ability to test the theory or to infer the theory from empirical data is therefore limited.

An example of current interest is "cosmic variance" in the predictions of temperature fluctuations in the cosmic background radiation. The observed pattern of temperature determines the correlation function $C(\theta)$ between the temperature fluctuations δT at two different directions $\vec{n}_1$ and $\vec{n}_2$ on the sky separated by an angle θ:

$$C(\theta) = \left\langle \frac{\delta T(\vec{n}_1)}{T} \; \frac{\delta T(\vec{n}_2)}{T} \right\rangle, \tag{8.1a}$$

where $\langle\cdot\rangle$ denotes an average over all directions $\vec{n}_1$ and $\vec{n}_2$ such that $\vec{n}_1 \cdot \vec{n}_2 = \cos\theta$. This correlation function can be conveniently expanded in spherical harmonics $P_\ell(\cos\theta)$:

$$C(\theta) = \sum_{\ell=0}^{\infty} \left(\frac{2\ell+1}{4\pi}\right) C_\ell P_\ell(\cos\theta). \tag{8.1b}$$

The coefficients C_ℓ so defined are the way the data from observations are usually quoted and are the objects of theoretical prediction.[12]

The probabilities of temperature fluctuations in the cosmic background radiation are predicted from a spectrum of fluctuations implied by the initial quantum state. The probabilities of these fluctuations are thus a detailed prediction of quantum cosmology that stem directly from the initial condition. They are not conditional probabilities requiring other information. The theory does not predict high probabilities for particular fluctuations in temperature at particular locations on the sky. Rather, it predicts distributed probabilities for these fluctuations (see e.g., [25]), or equivalently, the probabilities for various values of the coefficients C_ℓ. The expected value and the standard deviation of this distribution is shown in Figure 5. The width of the distribution is "cosmic variance."

We cannot test these probabilistic predictions for the cosmic background temperature fluctuations by measuring these fluctuations in a large number of identical cases. We have only one universe and only one set of observed temperature fluctuations! An observed distribution of the C_ℓ's inside this "cosmic variance" would be confirmation of the theory of the initial condition. An observed distribution outside it would be evidence against it. However, observations will not distinguish two theories of the initial condition whose "cosmic variance" both surround the observed distribution. Thus, for such probabilistic predictions we are inevitably limited in our ability to test a theory of the initial condition.

More generally, as mentioned above, a theory of the initial condition can be tested only through predictions whose probabilities are so near certain that we would reject the theory if they were not observed. The sets of histories that lead to near certain predictions are just a small set of those for which probabilities are predicted.

There are limits, therefore, to the process of inferring the initial state of the universe from observation. If $C_{\rm obs}$ is the operator describing the entirety of our collective observations, then strictly speaking all we can conclude about an initial state $|\Psi\rangle$ is that it is not such that

$$C_{\rm obs}|\Psi\rangle = 0. \tag{7.3}$$

This is not much of a restriction. For example, suppose that the projection $P_{\rm pres.\ data}$ represents all our present data, including our records of the

---

[12]See, e.g., [24] for a detailed review.

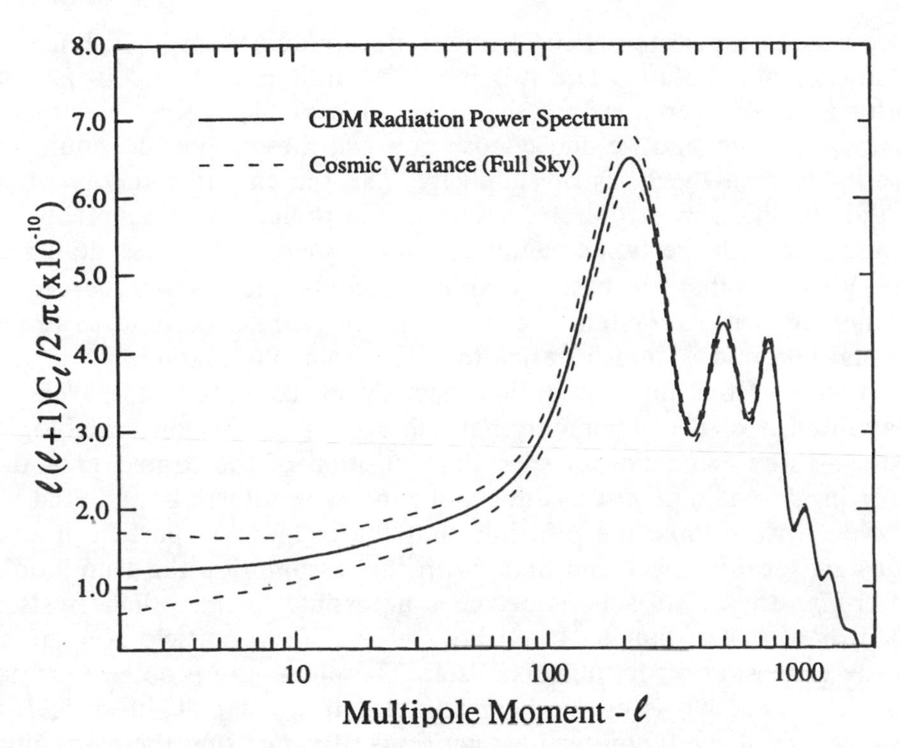

Figure 5. Cosmic variance. The heavy line on this figure shows the expected value of the multipole moments of the two-point correlation function defined by Eq. (8.1) for temperature fluctuations in the cosmic background radiation as predicted from the probabilities of these fluctuations arising from a simple theory of the universe's initial condition. The dotted lines indicate the standard deviation of the predicted distribution, called the "cosmic variance." Observations of our single universe yield the correlation function and one particular distribution of observed multipole moments. These observations will not distinguish two theories of the initial condition whose "cosmic variance" both surround the observed distribution. [Graph by J. Gundersen.]

past history. It is not possible on the basis of either present or future observations to distinguish an initial state $|\Psi\rangle$ from that defined by

$$|\Psi'\rangle = \frac{P_{\text{pres. data}}|\Psi\rangle}{\|P_{\text{pres. data}}|\Psi\rangle\|}. \tag{7.4}$$

Retrodictions of the past from present data and $|\Psi\rangle$ could differ greatly from those from the same data and $|\Psi'\rangle$.[13] But retrodictions are not

[13]Unlike classical physics, where the past can be retrodicted from sufficiently

accessible to experimental check and therefore do not distinguish the two candidate initial states. The two initial conditions, $|\Psi\rangle$ and $|\Psi'\rangle$, could differ greatly in complexity if the description of $|\Psi\rangle$ is short but that of $P_{\text{pres. data}}$ is long, and we may choose between these physically equivalent possibilities on the basis of simplicity. The search for a theory of the initial condition must therefore rely on the principles of simplicity and connection with the fundamental dynamical theory in an essential way.

Why is it that the basic dynamical theory—the Hamiltonian of the elementary particle system—seems so much more accessible to experimental test and so much easier to infer from observational data than the theory of the initial condition? Strictly speaking, it is not. Were the Hamiltonian of the elementary particle system to vary on cosmological scales—to be a function of spacetime position of the form $H(x)$—then inferring H would be just as difficult a process as inferring the initial $|\Psi\rangle$. However, we *assume* the principle that the elementary particle interactions are local in space and time. With that assumption the Hamiltonian describing these interactions becomes accessible to many local tests on all sorts of scales, ranging from those accessible in particle accelerators to the expansion of the universe itself. Therefore, the problem of inferring the initial $|\Psi\rangle$ is not so very different from that of inferring H in making use of the theoretical assumptions. It is just that the assumption of locality is so well adapted to the quasiclassical realm of familiar experience that many more tests can be devised on small scales of a theory of H than we are ever likely to find of a theory of $|\Psi\rangle$ on cosmological scales.

IX. Conclusions: The Necessity of Limits to Scientific Knowledge

If the world is complex and the laws of nature are simple, then there are inevitable limits to science. Not everything that is observed can be predicted; only certain regularities of those observations can be predicted. Even given a theory, computational intractability or observational difficulty may limit our ability to predict. In a world of finite observations, there are inevitable limits to our ability to discriminate between different theories by the process of induction and experimentation.

---

precise present data alone, retrodiction in quantum theory requires present data *and* the initial condition of the system in question. For further discussion see, e.g., [5], Section II.3.1.

Quantum cosmology—the most general context for prediction in science—exhibits examples of all three kinds of limits to scientific knowledge. There are only a very few predictions of useful probabilities that are conditioned solely on simple theories of dynamics and the universe's initial condition. There is a far richer variety of useful probabilities conditioned on further empirical data that are the basis for most of the predictions in science. There are some indications that the "no boundary" initial wave function is non-computable in the technical sense of yielding non-computable numbers. That does not limit our ability to extract predictions from the theory, in principle, but may be an indication that predictions sensitive to the topological structure of spacetime on small scales could be conceptually challenging to compute. Finally, it is possible to exhibit different theories of the initial condition with identical present and future predictions that can only be discriminated by an appeal to principles of simplicity and harmony with basic dynamical laws.

We should not conclude a discussion of limits in science without mentioning that science is useful *because* of its limits. Complex adaptive systems are successful in evolution and individual behavior because they identify and exploit the regularities that the universe exhibits. Scientific theories predict what these regularities are and explain their origin. Theories can be used to estimate how tractable these predictions are to compute or how practical they are to measure. By comparing different theories induced from the same data, an idea can be gained of the reliability of our predictions. The existence of limits of the kind we have discussed therefore does not represent a failure of the scientific enterprise. Limits are inherent in the nature of that enterprise, and their demarcation is an important scientific question.

Acknowledgments

The author is grateful to M. Gell-Mann for discussions of a number of the issues discussed in Section VIII over many years. Thanks are also due to H. Morowitz and R. Shepard for instruction in certain aspects of biology and psychology respectively, and to J. Gundersen for supplying several of the figures. This work was supported in part by the US National Science Foundation under grants PHY90–08502, PHY95–075065, and PHY94–07194.

References

[1] Bohr, N. *Atomic Physics and the Description of Nature.* Cambridge University Press, Cambridge, UK (1934).

[2] Gell-Mann, M. *The Quark and the Jaguar.* W. H. Freeman, New York (1994).

[3] J. Halliwell, in *Quantum Cosmology and Baby Universes: Proceedings of the 1989 Jerusalem Winter School for Theoretical Physics*, S. Coleman, J.B. Hartle, T. Piran, and S. Weinberg, eds., World Scientific, Singapore (1991) pp. 65–157.

[4] Woo, C. H. *Phys. Rev. D39* (1989), 3174; *Found. Phys. 19* (1989), 57, and in *Complexity, Entropy, and the Physics of Information,* W. Zurek, ed. Addison Wesley, Reading, MA (1990).

[5] Hartle, J. B. in *Quantum Cosmology and Baby Universes: Proceedings of the 1989 Jerusalem Winter School for Theoretical Physics,* S. Coleman, J.B. Hartle, T. Piran, and S. Weinberg, eds., World Scientific, Singapore (1991) pp. 65–157.

[6] Hartle, J. B. in *Directions in General Relativity, Volume 1: A Symposium and Collection of Essays in honor of Professor Charles W. Misner's 60th Birthday,* by B.-L. Hu, M.P. Ryan, and C.V. Vishveshwara, eds., Cambridge University Press, Cambridge (1993).

[7] Gell-Mann, M. and J. B. Hartle. *Phys. Rev.,* D47 (1993), 3345.

[8] Griffiths, R. *J. Stat. Phys.,* 36 (1984), 219.

[9] Omnès, R. *J. Stat. Phys.,* 53 (1988), 893; *ibid,* 53 (1988), 933; *ibid,* 53 (1988), 957; *ibid,* 57 (1989), 357; *Rev. Mod. Phys.,* 64 (1992), 339; *Interpretation of Quantum Mechanics.* Princeton University Press, Princeton, (1994).

[10] Gell-Mann, M. and J.B. Hartle, in *Complexity, Entropy, and the Physics of Information, SFI Studies in the Sciences of Complexity,* W. Zurek, ed. Addison Wesley, Reading, MA, 1990 or in *Proceedings of the 3rd International Symposium on the Foundations of Quantum Mechanics in the Light of New Technology,* S. Kobayashi, H. Ezawa, Y. Murayama, and S. Nomura, eds., Physical Society of Japan, Tokyo (1990).

[11] Joos, E. and Zeh, H.D., *Zeit. Phys.,* B59 (1985), 223.

[12] Zurek, W., *Phys. Rev.,* D24 (1981), 1516; *ibid.,* 26 (1982), 1862.

[13] Hartle, J. B. in *Proceedings of the Cornelius Lanczos International Centenary Conference,* North Carolina State University, December 1992,

J. D. Brown, M. T. Chu, D. C. Ellison, and R. J. Plemmons, eds. SIAM, Philadelphia, (1994).

[14] Fraser, C. M. et. al. *Science,* 270 (1995), 397.

[15] Shepard, R. N. *Psychonomic Bulletin & Review,* 1 (1994), 2 and *World and Mind* (to be published).

[16] Landauer, R. *IEEE Spectrum,* 4 (1967), 105; see also, *Proceedings of the 3rd International Symposium on Foundations of Quantum Mechanics in the Light of New Technology,* S. Kobayashi, H. Ezawa, Y. Murayama, and S. Nomura, eds., Physical Society of Japan, Tokyo (1990); and *Physics Today,* 44 (1990), 23.

[17] Hartle, J. B. and S. W. Hawking. *Phys. Rev.,* D28 (1983), 2960.

[18] Hartle, J. B. *J. Math. Phys.,* 26 (1985), 804.

[19] Rubenstein, H. *The Solution to the Recognition Problem for* S^3, unpublished lectures, Haifa, Israel (1992).

[20] Thompson, A. *Math. Res. Lett.,* 1 (1994), 613.

[21] Haken, W. in *Word Problems,* W. W. Boone, F. B. Cannonito and R. C. Lyndon, eds., North Holland, Amsterdam, (1973).

[22] Hartle, J. B. *Class. & Quant. Grav.,* 2 (1985), 707.

[23] Schleich, K. and D. Witt, *Nucl. Phys.,* 402 (1993), 411; *ibid.,* 402 (1993), 469.

[24] Geroch, R. and J. B. Hartle. *Found. Phys.,* 16 (1986), 533.

[25] Bond, J. R. in *Cosmology and Large Scale Structure, Proceedings of the Les Houches School, Session LX, August 1993,* R. Schaefer, ed., Elsevier Science Publishers, Amsterdam (1995).

[26] Halliwell, J. and S. W. Hawking. *Phys. Rev. D,* 31 (1985), 1777.

Chapter 6

STRUCTURING REALITY: THE ROLE OF LIMITS

PIET HUT

I. Introduction

Is scientific knowledge in principle limited? Are there things that we will never know about? Or are there perhaps aspects of reality that we may gain knowledge of, but in a form that cannot be translated into a scientific type of knowledge?

These are bold questions, and not the type of questions that would normally come up during a scientific conference. But in a workshop on *limits to scientific knowledge,* they are unavoidable.

It would be far easier to answer these questions if they were framed in a more relative way. Is scientific knowledge limited in practice? Of course it is: there are economic and political as well as ethical limits to the sort of explorations scientists could be expected to carry out. Are there areas of human knowledge that have not (yet?) been incorporated in scientific descriptions? Certainly, and we do not even have to argue about the status of aesthetics or deep human emotions. Many simple tasks of pattern recognition that even unskilled humans can do instantaneously and without special effort are still beyond the reach of the fastest computer with the fanciest software. It is clear that much of our tacit knowledge of day-to-day circumstances remains largely unexplored in scientific terms.

Scientific research is framed in terms of contemporary limits, and indeed is a continuous battle with limitations on all levels. The expressions most often used already tell the story: work at the frontier of science, research at the cutting edge, struggles to extend knowledge beyond their current limits. Usually these activities take up all the energy and attention of the working scientist. At our workshop, however, we are not so much concerned with the next break-through, or even the whole process of expanding scientific knowledge. Rather, we are interested in the scope and structure as such of the limits that scientists encounter. Hence the bold questions raised above.

Here is a tentative answer: perhaps there are no limits to knowledge, scientific or otherwise. If all limits were only relative to their particular context, none of them would have absolute or ultimate validity. This would imply that knowledge is open-ended, and unlimited in principle.

The Nature and Status of Knowledge

In a nutshell, here are the problem, motivation, and suggested solution. The *problem* at the core of any discussion of limits to scientific knowledge is the question of what knowledge is. The *motivation* for the specific angle of my inquiry into this problem is that I want to avoid a flattening-out reduction, materialist or otherwise, of knowledge—in all its forms, scientific knowledge as well as tacit understanding, meaning and sense, in short, any form of experience. The *solution* I suggest is to explore a world view in which knowledge is not secondary with respect to space and time, neither bluntly (as an epiphenomenon) or subtly (as a supervenient property). Rather, I would like to trace out some of the consequences of a world view in which knowledge is just as primitive an aspect of reality as are space and time.

Together with a tentative answer, here is a tentative model: perhaps knowledgc is an integral part of reality, as fundamental as space and time. Space is pervasive, and so is time, albeit in a very different way. Time is not embedded in space, and cannot be localized; it is really on a par with space. We could entertain the notion that knowledge, too, cannot be reduced to patterns in space and time, but is in some sense orthogonal to both.

Just as a speeding bullet does not "have" more time than a resting rock, perhaps a human brain or an encyclopedia does not "contain" more knowledge than a rock or a flower. Sure, the motion of a bullet expresses time a lot more vividly than a resting rock does. And sure, a book or a brain may well express knowledge in ways that strike us as being more vivid and more easily recognizable. Time can *express* itself in space in many ways, and so can knowledge. The question is to what extent either of them can be *reduced* to space.

Without clocks, we could not measure time accurately. But this does not mean that clocks somehow "produce" time. Any and every event can be seen as a witness to time, clocks or no clocks. Without human brains, no human knowledge. But this does not necessarily mean that brains somehow "produce" knowledge.

The space and matter of a speeding bullet do not generate time. This means that time is not reducible to either space or matter, and has to be invoked as a separate, equally primordial ingredient, in order to describe the situation. True, relativity theory has shown how space and time can partly transform into each other, since they both can be seen as aspects of a more general spacetime. But fundamental distinctions between space and time remain, not the least of which is the directedness of time (the "arrow of time"), so unlike the more symmetric properties of space. In addition, a notion of spacetime deals with space and time on an objectified level, and it is not at all clear how subjective experience fits into this (the problem of the "moving now": in objective spacetime, each time has the same status, whereas for us there is always only one subjective now).

Can space and time, together with matter in space and time, somehow generate experience? I would guess that most scientists would answer this question affirmatively. Experience is seen as a something that is "produced" in a brain, as an epiphenomenon, or an emergent property, or a supervenient aspect. I have two problems with such views.

First, I cannot even begin to see how a collection of molecules, when dancing in sufficiently complex patterns, could "give rise" to experience. Not only is molecular dynamics qualitatively different from experienced phenomena, it is experience itself that forms the condition of possibility for the arising of phenomena, as well as the arising of an understanding of molecular dynamics.

Secondly, and in many ways more importantly, I am worried about the implications of such type of explanations. When experience is "reduced" to complex interactions of molecules, there seems to be no alternative other than to continue to play this reductionist game. For example, we can then view all of ethics and aesthetics as properties of the brain, properties with significant survival value, selected by natural evolution. As a result, natural sciences in general, and physics in particular, are what ultimately have value, since they describe the only "really real" substratum of reality, the material world. All other "values", including the ones we hold personally most dear, are only derived values, tricks with which evolution has endowed us in order to be more successful in our never-ending struggles of competition.

At the same time, I am struck by the force of the arguments. Indeed, science has been incredibly successful, and indeed, walls of prejudice have crumbled, over and over again. Organic matter turns out to be as mundane as inorganic matter; no wall could be found separating the two. Now that we begin to have the blueprints in hand of some of the simplest organisms, it would seem only a matter of time before we can assemble the first living cell from scratch, from non-living material. When this happens, another wall will come down, and conceptually it has come down already, in anticipation. Why not expect that matter and mind are equally continuous as inorganic and organic chemistry, and as non-living and living matter?

These are forceful arguments. And no piecemeal attempt at denying or ignoring them will hold up to serious scrutiny. The only alternative I can see is a radical one: rather than tinkering with the explanatory structure of events in reality, seen as a stage made up of space and time, I suggest that we extend our view of the stage itself. If the stage includes a third element, on a par with the other two, space and time, then experience could be seen as irreducible to space and time (and matter, for that matter; see below for a connection with modern physics, and its description of matter as waves in fields that permeate space and time).

In other words, the standard scientific view starts off by introducing the existence of not-knowing (not-experiencing and not-conscious) building blocks. When put together in sufficient number and complexity, this view then insists that knowledge somehow emerges out of those entities. Unwittingly, such a view filters down knowledge, degrades it, tames and enslaves it unnecessarily. Rather than letting knowledge bloom and appreciating it to its depth, such a view, in my mind, starts off by approaching knowledge with a built-in set of limitations [1].

The question of the role of knowledge will be taken up again in the last section of this paper. First, however, we will have a look at some particular forms of limits, in the section directly below. The second section will focus on various aspects of objectivity, the cornerstone of science. The third section explores different aspects of limits in its various guises, as edges and boundaries.

II. Three Explorations of Limits

To begin our investigation of limits, let us start with three journeys, in each of which we will explore a particular limit to our knowledge. The first one will take us to smaller and smaller structures in space. The second journey will take place in time, and will take us back to the beginning of the Universe. The third will not imply any travel in either time or space, but instead will shift our focus away from a purely objective description of the world in order to include the role of the observing and participating subject.

Each of these sallies will lead to a paradox. Whether we will try to push the limits of space, or of time, or of experience, in each of these cases we will be presented with a surprise: after an initial process of narrowing, we will suddenly find ourselves in a realm far wider than the one we started out with.

An Exploration in Space

Let us take a very brief tour, from our human length scale to smaller and smaller scales, in order to get some sense of where scientific investigations have led us, this last century.

We can start with a physical object, a tree for example. What do we find when we analyze its inside? Depending on the scale on which we look, we find either wood and fluids, or wood cells, or molecules and atoms, or subatomic particles, or a seething sea of vacuum fluctuations. Although all of these structures inhabit the same location in space, and somehow together make up the tree, they also seem to inhabit separate worlds: when we go up or down one level of description, the previous level seems to have disappeared beyond recognition.

To be specific, let us choose just one scale, and make ourselves more familiar with it. Let us go to the first subatomic level of description, the one in which we view each individual atom as made up of a nucleus, surrounded by a cloud of electrons. The nucleus itself is made up of protons and neutrons, two types of particles which differ mainly in their electric charge: protons have a positive charge, whereas neutrons are electrically neutral, hence their names (electrons carry a negative charge).

The first thing that is striking on this level is the overwhelming emptiness of the subatomic world. The size of the nucleus is tiny compared to that of the atom, even though most of its mass is concentrated in the nucleus. While the mass of all the encircling electrons makes up only a quarter of a tenth of a percent of the mass of the atom, the diameter of the nucleus is a factor one hundred thousand times smaller than that of the electron cloud. In more striking terms, the nucleus inside an atom, even though it contains nearly all its mass, is like a flea inside a cathedral.

There is a nice anecdote about a popular talk given by Lord Rutherford, who had discovered the overwhelming emptiness of the atom around the turn of the century. When he described his model of the atom, a man in the audience objected saying that this theoretical idea was clearly refuted by the fact that walking into an iron beam was far from an experience of emptiness. Clearly, in such a case the beam is massively present, no matter what an atomic physicist may say. Rutherford's answer was short and simple, something along the lines of "the reason, dear Sir, of your discomfort in walking into an iron beam stems from the fact that your head is even more empty than the iron beam!"

More recently, we have begun to probe the inner structure of protons and neutrons, and found them in turn to be built up out of other particles, such as quarks and gluons. Further investigation may find yet other layers of structure, but let us stay for a moment with a description on the level of neutrons, protons, and electrons as the main constituents of matter. What we find at this level is a very pretty picture indeed.

Isn't it marvelous, to realize such a high degree of unity, underlying the vast diversity of materials and processes in our physical world? Whether glass or stone, wood or water, smoke or mud, all these materials are manifestations of the same three types of building blocks. The differences between different materials are simply consequences of the different configurations of the electrons, neutrons, and protons. These three particles are all that is needed to build up the nearly hundred different types of atoms found in nature. These atoms, in turn, form the components from which a vastly larger – indeed unlimited – variety of molecules can be put together. But for all the differences in appearance between materials, science tells us that we are dealing with the *same* basic constituents. The different properties are consequences only of differences in the *configurations* of the building blocks.

In a very real sense, then, one could assign the properties of different materials not to the protons, neutrons, and electrons they are built out of, but rather to the spatial arrangement of those building blocks. A surprising notion, that all we see around us in its bewildering variety is attributable to spatial, rather than material properties. Sticks and stones and bricks and bones are different by the different way in which the same constituents use space in a different way. So much for our notions of material substance.

The description given here is of course sketchy. An accurate physics description of the model of the atom in terms of protons, neutrons and electrons is rather complicated. It takes into account the quantum mechanical nature of the particles, with its intrinsic limitations to the accuracy and meaning of measurements. It also describes the force fields and potential energy fields between the constituents of the atom. Usually, the different material properties of different atoms and molecules are traced back to differences in these fields. However, since these fields themselves are characterized by the distances and angles between the spatial distributions of the neutrons, protons, and electrons, it seems equally valid to assign those material properties to the space in which the three types of particles are embedded.

A more accurate way to describe this curious state of affairs is to treat matter and space together, in a description of material particles as excited states, waves in a field which by itself pervades all of space. In quantum field theory, the language of modern particle physics, fields are always present, and the presence or absence of a particle is comparable to the presence or absence of a wave in the ocean. In this analogy, it is the interplay of the water, only indirectly visible through the behavior of the waves, which determines the properties of matter.

While we talked about distinct electrons above, as one of the types of building block for matter, this more accurate picture views all the different electrons in the Universe as being only differently localized excitations in the same electron field, different waves in the same ocean. And we may conceivably discover the different fields describing quarks, electrons, and other "elementary" particles to be only facets or projections of one underlying field. At present, there is a bewildering variety of candidates for more unified theories, some of which even include gravity.

Independent of the details of future developments, however, it is clear that a search for smaller and smaller constituents of matter leads us to a paradoxical situation. At first, a piece of matter can be subdivided into smaller and smaller pieces, down to molecules and atoms. But then, suddenly, the subatomic particles that we find "inside" an atom turn out to be not really "inside." Yes, as an excitation of a field they are largely localized inside. But their very presence requires a corresponding field that fills the whole of the Universe.

It is as if the smallest peeping hole suddenly has given us a view of the largest scales. By trying to brush a crumb from the table, we run into a surprise: what looked like a crumb turns out to be a pattern that is woven into the table cloth. We find ourselves pulling the whole table cloth with us, together with everything that seemed to rest on it.

An Exploration in Time

Let us now shift from space to time, and let us embark on a quick survey of the view science has presented us of our own history, as well as the history of our Universe. The story runs as follows.

The Big Bang started off as a very hot and very dense soup of elementary particles. While rapidly expanding and cooling, some fraction of these particles were converted into hydrogen and helium during the first three minutes of the history of our Universe, following the Big Bang. Then, some time during the first few billions years, much of the primordial mixture of hydrogen and helium gas started to clump, here and there falling together under its own gravity. In this way galaxies were formed, and around this time stars started to form as well, through gravitational contraction on much smaller scales, deep inside the galaxies or proto-galaxies.

Around some of those stars a small fraction of left-over material did not make it all the way in, and later underwent subsequent gravitational clumping to form even smaller bodies circling the parent star: planets, asteroids, and comets. Our Earth is one such planet, and was thus formed as a by-product of the formation of our Sun, a later-generation star, formed several billions years after our Galaxy (visible for us in the form of the Milky Way) was first assembled.

A billion years or so after the Earth was formed, a random interplay of macromolecules led to self-reproducing chemical reactions complex enough to form their own tiny laboratories: the first primitive cells. Inside the first defensive suits, in the form of the cells' walls, ongoing chemical experimentation and natural selection worked hand in hand, resulting in further differentiation. This led to the appearance of multicellular organisms, and especially in the last half billion years, to an explosion of diversity of plant and animal life in the sea and on the land.

A few million years ago, homo sapiens appeared as one more product of this evolutionary Monte Carlo game. For a long time we lived as hunters and gatherers, until more and more of us began to settle down some ten thousand years ago. And here we are, a few hundred generations later. We can trace the shapes of our ideas to a few thousands years of civilization. We can trace the shapes of our genes back to the beginning of mankind, and the composition of our DNA's building blocks back to a much earlier origin, billions of years ago. On an even more elementary level, we are literally a form of star dust: the chemical elements making up our body are ashes of nuclear reactions that took place in previous generations of stars, that were born and died before our Sun was formed.

These ashes, in the form of atomic nuclei, are in turn built up out of protons and neutrons, particles that have been around much longer. But they, too, were once assembled from lighter constituents, quarks, at a very early time in the history of the Universe. This happened less than a millisecond after the Big Bang started off. And what about those quarks? Presumably they, too, can be seen as products of even earlier transmutations, although the notion of one type of particle constituting, i.e., forming the building blocks for, another type of particle may well break down here (already in the case of quarks, the analogy is less than perfect).

At present, we can only speculate about what happened at much earlier times, when the Universe was so hot that a typical particle in this primordial soup was more energetic than the highest-energy particles that we can produce in our most advanced accelerators. There are clear hints that the behavior of matter is distinctly different at sufficiently high energies, corresponding to an age of the Universe of a mere 10^{-43} seconds or less, when gravitational interactions were at least as strong as other types of interactions between particles.

One speculation is that at times earlier than 10^{-43} seconds, the properties of time and space are radically different from those we have been able to probe so far in the laboratory. Perhaps even notions such as distance and duration are no longer valid. Could it be that at the "time" of the Big Bang, or soon thereafter, a particular trajectory in time was set up, from within a "timeless" state that somehow "preceded" our present cycle of expansion (and possible contraction) of the Universe?

Whatever the answer may be, a search for earlier and earlier phases of the Universe is likely to lead us into a paradoxical situation. The paradox here stems from the fact that the Universe seems to have started at a very definite point in time. This is a startling conclusion. Only a century ago, the leading scientific and philosophic opinion favored a Universe that had been around forever. It was considered all too anthropomorphic to think that the Universe would share such a property of human beings as being born a finite time ago. It seemed clear that an eternally existing Universe was far more attractive.

However, observations told us otherwise. Given the observed expansion of the Universe, general relativity tells us that the Universe has started not too long ago, a mere ten or fifteen billion years before the present. The Universe as a whole is thus not that much older than our home planet, with an age of four and a half billion years.

In a search for earlier and earlier times, we seem to reach a point where time began, and attempts to reach for even "earlier" times may well lead us into a literally time-less realm, or at least time-as-we-know-it-less realm. As was the case in our exploration of space, an exploration of time, too, may turn out to open up, from a descent into smaller and smaller time intervals, out onto a vista beyond time.

An Exploration in Experience

After having traveled in space and in time, what other direction is there left to travel in, exploring the limits of our knowledge of reality? From the viewpoint of natural science, none, or at least none that has the same level of fundamental significance. But natural science itself is the result of our taking a particular stance towards our experience, in which we filter out certain parts in order to arrive at an objective view of nature. And it is easy to overlook this filtering operation: since we have been so very successful in the scientific enterprise, the temptation is strong to try to stuff everything back into this limited domain.

This sleight of hand works as follows. We take our experience as given, dismiss large fractions of it as (temporarily at least) irrelevant, and use the remainder to build up an objective scientific world view. It is this world view that we have briefly reconnoitered in the previous two sections. And then, at the end of the day, we conclude that all of our experience must be somehow "produced" by human brains, as a result of complex interplays of large numbers of molecules. In other words, having filtered out most of our experience, and having transformed the remainder in a highly abstract manner, we give our final fiat to the belief that somehow all of experience can be recovered nonetheless, as part of a single objective reality that includes our subjective experience as the output of complex computers called brains.

From Bishop Berkeley onwards, critical voices have been heard, protesting against this glib type of reductionism, but by and large the world view summarized above is the one we use as our standard reference. After all, there seem to be two good reasons to do so. First, there is the fact that science has been so extraordinarily successful, for better or worse. Second, there is the question of what else we can do. What alternative do we have, apart from buying into a scientific world view, if we do not want to wind up with romantic speculation or religious beliefs?

Well, let us explore and see for ourselves. Let us try to drop all forms of dogma and prejudice, including the ones that tell us that our identity is to be found as a highly complex and brittle collection of material particles in space and time. One of these dogmas is that we are completely insignificant with respect to the far vaster Universe we live in, the result of a large variety of random processes, initiated with the Big Bang. Whether these dogmas may or may not be correct is not the point here. The challenge is to put them on hold for a while, and to try to look at the world with fresh eyes.

In other words, the challenge is to make a shift in experience, a shift that is completely independent of the more familiar shifts in space and time, as exemplified by the processes of motion and aging that we are all involved in. The shift in experience I have in mind is the one advocated by the German philosopher Edmund Husserl (1859–1938).

I consider Husserl to be one of the foremost experimental philosophers, perhaps the first philosopher in the age of modern science who managed to carry over the spirit of the method of experimental physics into philosophical research. Husserl introduced a specific experimental method, the *epoche* (from the Greek ϵπoχη for "suspense of judgment"). He advocated a kind of experimentation in which we just pay attention to what appears, while dropping our unquestioned allegiance to the world view we grow up with.

At first, it may seem to be very strange to "put the world on hold," to drop any belief in an objective reality as the prior and only "real" form of reality. But there is nothing magic or special in making this shift. It is only the result of the shift that is remarkable, a form of amazement and wonder. In fact, reactions of such a type are the touchstone to check whether a shift really has been made, or whether an attempt to "put the world on hold" has only been an intellectual game.

Many poets and novelists have testified to such a shift, a dramatic change in experience, away from a belief in a solid world in which we are anchored, and towards a completely open experience of the world as bottomless. Several philosophers, too, have given us an inkling of their experience along these lines. The problem is, in our culture, that poets typically give their experiential report without any theory, and philosophers tend to give only their theoretical reflections while glossing over the experiential component that undoubtedly underpins their theoretical moves towards more open interpretations of reality.

One philosopher who was unusually open and honest about both the importance and personal impact of experiential involvement in philosophy was William James. It was partly through inspiration he received from James, that Husserl made his far more detailed attempts to combine experimental and theoretical philosophy.

When reading Husserl, one is struck by the sense of honest amazement that is conveyed. An amazement about the way we make sense of the world, and a deep sense of surprise about sense, something we find everywhere but something we cannot catch. Like space, like time, sense is for us what water is for a fish. Our lives are embedded in it, given by it, irremovably linked to and through it. (Note: I use "sense" here in its aspect of "meaning," and I do not imply any connection with "sense experience"—the word sense conveys more of a grasp of something than the somewhat abstract word meaning).

Sure, we can interpret our world as a world of things. But what is a thing? When we look carefully, then we find that what we considered to be an object appears in our consciousness as a bundle of meanings, draped around sense impressions that are far, far less complete and filled in and filled up than the "real thing" we feel to be present, three-dimensionally, continuous in time. What then remains of the solidity of the object? It is recognized in its givenness for us through the *sense* of solidity we have. Its continuity? This follows from our *sense* of continuity and identity. Its reality? Nothing but a *sense* of reality. The indubitability of its reality? The only thing we have a real handle on is our *sense* of indubitability of its reality [2].

Husserl himself did give ample indications of the fact that for him the *epoche* was a way of life, and towards the end of his life he described it as a "complete personal transformation, comparable in the beginning to a religious conversion" [3]. And indeed, recognizing that we live in a world of sense is a shocking experience. Not only do we find the *world* to be dissolved in sense, upon close inspection, but we find that *we ourselves,* too, are known to ourselves only as complex forms of sense. This shift towards sense is far from an armchair philosophy consideration, it takes place in the laboratory of our life, as is nicely expressed by Harvey [4]:

> Husserl's procedural techniques for inducing the "shift" are an attempt to articulate a certain strange experience that has happened to philosophers, to artists and poets, and perhaps to everyone save the hopelessly sane, now and again throughout their personal history. This strange experience is the experience of the strangeness of experience, and of the world. And this strangeness is nothing more (nor less) than the act of *seeing through* the sedimented meanings that one inherits and develops, and that structure one's world.

One important point has to be emphasized here. When we take a turn towards experience as primal, rather than derived, it is not at all clear "whose" experience we are talking about. We normally assign experience to ourselves, to a person with a certain identity, playing all kinds of roles. But aren't all these aspects given as part of experience? In this context, it is interesting to read the Japanese philosopher Nishida's reaction [5]: "Over time I came to realize that it is not that experience exists because there is an individual, but that an individual exists because there is experience." After all, we can only know objects in the presence of a subject, just as we know ourselves as subject only through our interaction with objects, be they thoughts or things or other forms of appearance. And since it all leads back to experience, it does seem reasonable to start there, and to consider both subject and object to be attributes of experience, rather than the other way around.

So it would seem that the whole world dissolves into a world of experience. But that is not quite right. If we put on hold the notion of "world," "person," and "self," then we cannot label what happens as being the experience of a self. Yes, of course, things still happen, but we can refrain from calling it experience. What then to call it? How about calling it appearance. It is clear that something is going on. Something happens. Something appears. What appears? Appearance appears. That's all. We can say more, but anything we say is likely to let us stray away from appearance, into secondary considerations.

We have thus arrived at a third paradox. We started with a vast objective world, and found ourselves as tiny specks in the physical Universe, as individual observers, living on the surface of a particular planet. But such a picture identifies the subject of experience with a particular object, the particular human body that the subject finds in the center of experience. From an objective point of view, such identification of course makes sense. But from a subjective point of view, the locus of experience is not to be found in my head, but everywhere around me. I am not locked up in my head, but I "am" in my experience. I live in the living world around me, as that world, as the only tangible part of that world, namely as the way in which it appears.

From this latter point of view, the move to ground experience in a brain leads to a similar surprise as the one we have seen before. Starting with space, trying to zoom in on a small part of it, we instead saw all of space reflected in the existence of a single electron. Starting with time, trying to trace it back, we found a starting point in the form of the Big Bang, and a hint of a realm beyond time. And now, starting with experience, trying to trace it back to its source, we again have a similar choice. Do we stop at the brain, waving a magic wand of "explanation," insisting that somehow experience oozes out of the complex interplay of molecules that makes up the brain? Or do we embrace the whole field of experience, including the experience of the physical universe, as a given, as equally fundamental aspects of reality as space and time?

The latter view, in my opinion, is far more promising than a hierarchical view in terms of mysterious "emerging" properties. It also seems to offer a common ground for dialogue between a far wider set of world views, European as well as non-European. In the following section, let us contrast this open view of the world, as a world of sense/meaning/knowledge, with the more closed view of the world as an objective realm that we are contemplating, as it were, from the outside, with the God's eye view of the ideal of a detached scientist.

III. A View from Afar

The desirability of developing a more global and less Eurocentric world view is an idea that is gaining in popularity. However, relatively little concrete development has taken place as yet in that direction. Especially in academic circles, a study of the philosophy and cultural values of non-Western countries is often still classified under "area studies," and seen as not on a par with the "real" philosophy that is considered to be the sole birthright of Europe.

Many Western philosophers are quick to point out the extent to which philosophical ideas in Asia are entangled with religion, be it Hindu or Buddhist or Taoist. Rarely do they pause to try to view themselves through reverse eyes, which would bring out the enormous entanglement between Christian dogma and both scientific and philosophic thinking, for better or worse, still very much so in even the latest postmodern way of phrasing the most pressing problems [6].

From a European point of view, we like to stress the tensions that have existed between science and the Christian church. But from a non-European view, the parallels must be far more obvious and striking. If we want to have a close look at the birth of modern science, we can compare seventeenth-century scientific thinking and seventeenth-century Christian thinking. In both cases, we see a tendency to view the world from an eternal vantage point. The Christian notion of a God transcending space and time gave a convenient model for a "God's eye" point of view, a way to look at reality from the outside, so to speak. This detached and stand-off-ish way of describing the world has led to a world view in terms of objectivity.

We are so much used to thinking in terms of an objective world, out there, mapped to high accuracy by physics, given once and for all. It seems almost impossible to put up for discussion the absoluteness of this objective world view, and the validity of taking such a stance. Questioning the ultimate status of objectivity is often interpreted as threatening the whole notion of a scientific attitude.

In order to study limits to scientific knowledge, we have to be aware of the limitations that are imposed on knowledge in order for it to qualify as scientific. Traditionally, any subjective form of knowledge was suspect in this regard, and generally considered "off limits" to science. True, any laboratory reading of any instrument ultimately is reported by individual subjects, perhaps at the end of a long chain of automatic measuring devices. In this sense, subjectivity lies at the basis of any form of scientific practice. The traditional justification of objectivity, as a construct on top of this subjective basis, was given by the ideas of verifiability and repeatability. These were invoked to safeguard us against unwarranted subjective interpretations and deviations. However, we have reached a point where we may have to rethink that whole strategy.

In physics, quantum mechanics has shown us that it is wrong to think that measurements can be arbitrarily repeated. Instead, the role of the observer has to be taken into account explicitly in a complete description of a quantum mechanical measurement. In cognitive science, a study of consciousness "from the outside," in objective terms, can only go so far, and leaves out the "insider view" of the one who experiences in subjective terms. And finally, when we look at the influence of science on society, we often see egocentric and non-caring attitudes being sold as being "scientific" and rational. Perhaps the strategy of limiting ourselves to objectivity, useful as it has been in the past, is doing us now more harm than good. But what are the alternatives?

The alternative I have in mind is something I would characterize as "freedom from identification." I will come back to that notion later. Let us first have a closer look at objectivity and its limitations.

A Particular Universe

An objective world view describes all concrete occurrences as instances of more general, universal laws. Each particular event that is scrutinized in the laboratory can be reproduced, in principle, by others. They can thus verify the objective validity of the universal laws that are inferred from particular measurements. The universal is what counts, and the particular is considered only in its incidental instrumental value. Laboratory objects (and often laboratory animals-as-objects as well) can be discarded after universal significance has been squeezed out of them. After studying one billiard ball, we can move on to the next, and the same is true of observations of clouds, or stars, or whole galaxies. The situation gets a little more problematic when we focus our gaze at the Universe as a whole. Notwithstanding its name, the Universe for us is not very universal at all, since we cannot trade it in for another one. We are literally stuck with the particular Universe we find ourselves living in.

There is an interesting parallel with the study of experience. In our third exploration we have found that experience can be viewed in two ways. As an object of study, we can observe someone else's experience indirectly, by studying that other person's behavior. And indeed, we can take a theoretical stance, and regard our own experience in a similar way, treating it as an object. But when we take a more directly experimental attitude, we find our own experience to be literally everywhere we look (or think, taste, etc.). We can see everything not only "in" experience, but in fact "as" experience. From an experiential point of view, it is impossible to take our own experience as an object, since we simply cannot take up a standpoint that is far enough removed. We cannot step outside our own experience.

This is then the parallel between the second and third exploration we have embarked upon earlier. It is a parallel between the inability to step out of *subjective* experience, and the inability to step out of our *objective* Universe, the one we live in, and the one we have traced back to its origin in the Big Bang. An analogy with mathematics may be helpful here, to shine more light on this comparison.

When we want to describe curvature, for example the curvature of a two-dimensional figure, we can look from the outside, and conclude that a sphere and a cylinder both are curved, while a plane is flat. But when we look from within the two-dimensional worlds, restricting ourselves to measurements that can be done from inside that world, we get a different result. The curvature of a sphere can be measured intrinsically, for example by drawing a triangle, and noticing that the sum of the angles is more than 180 degrees. However, the intrinsic curvature of a cylinder turns out to be zero. The results of any *local* measurements within the surface of a cylinder are indistinguishable from measurements made from within a plane. The only difference comes from the globally different boundary conditions: within the surface of a cylinder it is possible to return to the point of origin by walking in a straight line, something that cannot be done in a plane.

It may seem surprising that curvature can be measured from the inside of a surface, without ever leaving the surface in order to watch its curvature from an external view point. And indeed, in many cases there is no need to take an internal attitude. But there are situations in which no external view point is available. The most striking such example is given in cosmology. When we want to determine the curvature of the four-dimensional spacetime of the expanding Universe, we cannot just leave our spacetime, and travel in a fifth or sixth dimension in order to get a better view. As four-dimensional beings (three-dimensional *qua* space and one-dimensional *qua* time) we are given together with our four-dimensional spacetime, and there is literally no way out.

In the case of cosmology, it still is useful to construct models of the four-dimensional world that are embedded in higher dimensions. Similarly, it is useful to take only, say, two of the four dimensions of the world to illustrate some of its properties. For example, we can compare the expanding Universe with the two-dimensional surface of an expanding balloon embedded in a space with an additional third dimension. Such a model can roughly describe the kinematics of how the Universe expands (all points on the surface of the balloon receding from each other, without any surface point being the real center, the latter being located outside the two-dimensional surface altogether). But it is important to note here that the *external* curvature visualized here is the curvature of the *model.* As far as measurements go of the *real* Universe, such observations by necessity have to be *internal.*

For each individual human being, the challenge to make sense of the field of experience is akin to the challenge cosmologists face. Here is the parallel between observations in cosmology and in human experience. Each one forms a realm with, strictly speaking, no "outside," at least no outside in its own terms, connected in a continuum with the inside, sharing a common boundary. In both cases, we can only rely on internal measurements in order to understand our reality. If we then decide to construct models of reality, we can of course consider those models from an external point of view. An interesting paradox crops up here: such a model simultaneously *stands for* and is *part of* the Universe or the human experience it describes. Just as a model of our Universe is still part of the Universe, so are our models of experience all part of our experience.

When we now turn to an investigation of our own experience we find that each of us is a professional "cosmologist." We spend our working life, as well as the time in which we are sleeping, playing, or whatever, within the "Universe" of experience.

Experience is the foundation of our knowledge. It is the source, setting, and sphere of reference for all formulated assertions. It is a curious thing, however, that the knowledge we develop out of our experience, namely scientific conceptualization, leads to our providing an explanation of experience that in turn seems to provide a foundation for it. Starting from an internally felt perspective, we seem to bootstrap ourselves up and out to an external perspective. We conclude that the external empirical perspective, that is, the objective observational mode of experience is the more "real," the more fundamental, precisely in that it seems to provide a form of foundation that cannot be found within experience itself.

Something funny is going on here. That what we consider more "real" really is a form of abstraction, literally, when we analyze how reality is constructed in our experience. The whole notion of a single objective world, underlying all of our subjective experiences, is a handy and economic extrapolation, but one may wonder whether there is anything more to it than that. We will return to this question later.

Note that there is a potential problem here with the analogy between cosmology and human experience. Is a model of the Universe really "in" the Universe? Sure, the cosmology books are very much part of the physical Universe, but their message becomes meaningful as a model only in a mind. Is a mind "part of" the Universe? If the mind could somehow be located "in" a brain, the answer would be yes, but such a notion of "in" would seem rather too simplistic. The strong spatial sense of the experience-as-cosmology metaphor can thus be misleading when talking about experience, as if experience were a field of a very peculiar sort that is contained within the physical space-time Universe. This is precisely *not* what the metaphor tries to convey. To the extent that anything is within anything, physical space-time, that is, the dressing of really experienced space-time in the clothing of abstractly formulated scientific concepts, actually is "within" experience.

Stepping Out of Experience

Let us now take a less modern angle on this whole question of experience and of being "in." In talking about the Big Bang, we have used a physics view of the history of the world. Let us now have a look at the historical root of physics itself. How did we arrive at our current notion of an objective world view? In other words, how and when and where did we learn to "step out of experience," in order to look at the world in a disinterested, non-subjective way?

Our modern notion of an objective world, describable by science in terms of mechanical terms, is an unusually cold and remote one, as far as world views go. In most other cultures, a person constitutes a much more integral part of the world. Somewhere between heaven and Earth, in the grand scheme of things, humans play an important role. And not only human beings, also animals, plants, the whole interwoven tapestry of appearance can present a dimension that elicits wonder and awe.

By and large, we seem to have lost that dimension. Sometimes, in poetry for example, we may detect faint echos of this outlook on life, or better "inlook" in life. But when we close our poetry book, we again find ourselves as very small entities, almost lost in the vast time and space of the Universe, a product of an extremely lengthy chain of accidental occurrences which finally led to the aggregation of the material constituting our individual bodies.

Our bodies contain a nervous system complex enough to allow reflection on ourselves and our world. We can comfort ourselves perhaps by the fact that, small and insignificant as we may seem to be, we can to some extent grasp the whole Universe in our thought. But this by itself does little to overcome our sense of senselessness of the world. With all our progress in scientific thinking we seem to have lost our integral role in the world. And this loss is serious, and goes far beyond a romantic longing for the days of old. This loss has practical consequences, reflected as it is in the destructive way we treat our environment.

How did we get where we are now? What were the historical processes that have led to the thorough alienation we have arrived at, on many levels (with respect to nature as well as to neighbors, in the work place as well as in the political structures we form part of)? Of course, the development of natural science, over the last few hundred years, has played a significant role. But even earlier than that, there seems to have been a significant shift in attitude, in the way we take up our position towards the surrounding world.

Paintings from the Middle Ages, for example, up until the mid-fifteenth century, show a human world in which the viewer participates. A painting of a city shows a jumble of houses, piled on top of each other, each house showing itself as if you would be just walking in front of it. In contrast, in the late Middle Ages and early Renaissance, we see how the human role changes from a participant to an outsider. The invention of perspective puts the viewer away from the crowd, behind a window as it were, as if looking out from a separate world. According to Romanyshyn [7], this distance-creating move on the level of art prepared the ground for a similar move on the level of science.

This transformation of the human role, from player to spectator, has had profound consequences. Twentieth-century alienation may well have its roots in the invention of perspective in the fifteenth century as well as the founding of modern science in the seventeenth century. This is not to say that taking a detached view is something undesirable in itself. In order to face a complex problem, there is nothing wrong with trying to simplify it first, by reducing its full complexity to something that can be handled more easily. The problem lies in the danger of mistaking such a detached stance as providing a form of absolute reality rather than a tool.

Reductionism

Science would be inconceivable without an initial phase of reductionism. The notion that we approach something at first from a particular point of view, willingly neglecting other aspects, is natural. In fact, our embodiment forces us to look at objects from one perspective at a time, and it is only in our mind that we create an impression of a more "full" presence of a particular object, even though we construct (our experience of) that presence merely from the various views we take of it, from different angles. Strictly speaking, the experience of a three-dimensional object is a form of reconstruction, in our mind, based on more limited, "reduced" sense impressions. At least, this is the picture that neuroscience and cognitive science presents us. From this perspective, then, the felt three-dimensional presentation of a flower vase in front of us is a form of theoretical extrapolation, rather than an observation itself, although it is based on two-dimensional observations.

Traditionally, the particular approach to reductionism in science has been one of taking things apart: of an analysis of the whole in terms of its parts. The hope, here, was that a sufficient understanding of the parts, and their relations to each other, would enable us to see exactly in what way the whole is more than the sum of the parts; how exactly it can function in ways that none of the parts, by themselves, can.

In some ways, this hope has been fulfilled beyond expectation. At the beginning of this century, after a few hundred years of remarkable progress in physics, it was still completely unknown why different materials show different colors. Something as mundane as the greenness of grass or the brownness of rust simply had to be treated as a given, and lay beyond the explanation of physics.

Shortly afterwards, a breakthrough occurred. With our understanding of quantum mechanics came the ability to *calculate* properties of matter, such as color, that had previously eluded us. So finally, after a few centuries of treating "secondary properties" of matter, such as colors, as just that, as secondary citizens, it now was possible to include them within the arena of science. This was a triumph of reductionism: by putting the question of colors off-limits in the seventeenth century, the road was paved for progress in physics that ultimately would cover colors as well, in the twentieth century.

So it seemed that drastic filtering has paid off. But has it, ultimately? Does quantum mechanics really "explain" color? Yes, we may now be able to calculate how the wave length distribution of white light is affected by reflection off a particular type of material. But no, that is not quite the same as understanding why the experience of a red piece of matter as red has the particular qualitative "feel" that it does.

Let us illustrate this move towards world reconstruction with a simpler example, also involving color. Photography produces a picture of part of the world, in a two-dimensional projection. At first, black-and-white photography gave us an interesting, practically useful as well as aesthetically pleasing way to "paint with light." But of course a lot was missing: there was no depth, and there were no colors. In both ways, a photograph was a projection, a reduction in the range of openness to the rich variety of phenomenal features present in the real world.

Starting from a single such projection, there is no way to reconstruct the original scene. It is impossible to recreate colors by looking at a single black-and-white photograph. We can use cues and make educated guesses, but essential information is lost in the process of reduction. However, it is possible to do much better by using simultaneous projections along different angles (as with stereoscopic cameras) and in different colors (as in color photography, based as it is on the combination of several black-and-white exposures sensitive to different color bands).

One can imagine this process of parallel reduction and multi-faceted reconstruction to reach a state of perfection in which our visual perception can no longer distinguish the photographs (or rather its offspring, virtual reality) from "real" reality. In fact, all these forms of imaging can do more than simply recreate reality. Photographs can make visible to us forms of information that are not discernible normally, such as the ultraviolet light that some insects can see, or infrared light that allows us to see living bodies and working engines in the dark (well-known in military applications).

Can we say, in all these examples, that technology can succeed in recreating the world, partly (as in photographs) or more completely (as in a future form of virtual reality)? To some extent that is accurate, but it is important to realize one common factor between nature and a technological reproduction of nature. In both cases there is the human observer who is involved in the process of observing and who makes judgments as to similarities and differences between the two.

This special role of the observer is what makes the current state of the art in science so paradoxical. On the one hand, we are confident that we can model, in principle if not in practice, any type of physical process in daily life in considerable detail. Such simulations, either performed approximately by pen and paper or based on complex computer calculations, give us enormous power. We can design new materials, understand the properties of naturally given materials better, and in many ways we have become masters of our physical world.

But on the other hand, it is not clear that we are anywhere near a deep understanding of our own mind. We can measure correlations between objective brain states and our own subjective conscious experience in more and more detail, but it seems that an enormous gap still remains: the gap between a description of physical brain states and the direct personal experience of *qualia,* of conscious sensations such as experiencing the color red, a musical tone, or the smell of roses. And the same holds for thinking, feeling, remembering.

This gap is more than just a gap, it is more like an abyss between a physically objective description of the world, including the brain, and direct subjective experience. At first sight, this abyss seems to separate a scientific, third-person form of knowledge from the first-person knowledge that is the stuff of our life. The latter type of knowledge seems to be off-limits to scientific knowledge, and in that sense scientific knowledge forms a subset of the totality of our knowledge of reality.

Will scientific knowledge always remain a subset? Perhaps not. But before venturing out into speculations about the future, let us look a bit closer at the structure of the scientific framework, its subdivisions into disciplines, and the extent to which these disciplines are built upon each other.

The Hierarchy of the Sciences

Let us quickly review the typical way that scientists deal with the relation between the major scientific disciplines. Chemistry is an explanatory structure that is layered on top of physics. In principle, one might try to argue that any phenomenon in chemistry can be "explained" (calculated, checked) by physics, although in practice there are many obstacles to such a reduction. Apart from the fact that the computational problem may be intractable even for the fastest computers available now and in the foreseeable future, there is the more fundamental problem that higher levels of description require their own vocabulary which cannot be reduced to lower levels: "more is different," as Phil Anderson expressed it so eloquently, a quarter century ago [8]. Or in the words of Michael Polanyi, chemist-turned-sociologist [9]:

> Lower levels do not lack a bearing on higher levels; *they define the conditions of their success and account for their failures, but they cannot account for their success, for they cannot even define it.*

For a brief and clear introduction to this problem, starting with Husserl's view of the matter, see the article *Fundierung as a Logical Concept* by mathematician-philosopher Gian-Carlo Rota [10].

Even so, physics is generally considered to be the most fundamental science, with chemistry seen as more derived, and biology in turn layered on top of chemistry, and so on. Let us engage this picture for now, just to see where it leads us, and let us take psychology as the top layer as the study of the human mind (one could continue with sociology, economics, and other disciplines, but let us not make our picture more complicated, at this point).

Is there anything underneath physics, a "deeper" layer? Well, the natural candidate that would come to mind is mathematics. But as soon as we entertain the notion of "grounding" physics in mathematics, we face all kinds of problems. While chemistry seems to make use of the material substratum provided by physics, mathematics only provides pure form. We are unlikely, in this day and age, to go along with the Pythagoreans who held that all and everything under the sun (as well as the sun and stars themselves) could be reduced to numbers.

However, for all its inadequacies, let us briefly entertain this picture of math providing the foundation for physics providing the foundation for chemistry, etc., until we reach the study of the human mind, in psychology, founded on top of neurobiology, say. Let us look at this picture, and step back for a moment, trying to get a feel for the character of the interrelations between each pair of layers involved, as well as for the relations within each layer.

As soon as we ask the question of how each layer deals with itself, we come to a surprising conclusion. While all layers carry a wealth of information and thus have a lot of interesting stories to tell, they are mute when interrogated about themselves. Physics, for example, needs a form of metaphysics, something different from physics, in order to talk about physics. Physics talks about the physical world, and therefore by construction cannot talk about talking about the physical world. Physics and the physical world are two very different domains indeed. The same holds true for chemistry and biology and other disciplines in natural science. The surprising conclusion then is this: it seems that only the top and the bottom layer, psychology and mathematics, can be truly self-referential.

A model for mathematics can be constructed using pure mathematics. In other words, meta-mathematics can be mapped straight into mathematics, can be constructed as a form of mathematics itself. This is different for physics and chemistry, where the models are not part of their area of study (the physical world) but are couched in mathematical terms. But a model for mathematics, as pure form, is not different in structure from math itself, pure form from the outset.

Moving up the foundational layers from math to physics to chemistry to biochemistry and biology, we can never model any of those disciplines in terms of themselves. At best, we can make mappings, establishing isomorphisms between different systems that obey similar laws. But the laws themselves are of a different nature than the systems described. This holds true for physics, as well as for any "higher" level, until we reach psychology. In psychology we once more encounter a form of self-referentiality, encountered earlier only in mathematics: when studying consciousness, we use consciousness in order to make models for consciousness. In other words, being aware of being aware is itself again a form of awareness.

Clearly, reductionism in its classical form breaks down in the light of the problems that are inherent in this self-referentiality that we find at both extremes of the hierarchy. Yes, reductionism has played an important historical role in getting science started. But no, reductionism cannot be more than a first approximation. Once the engine of science has been jump started with a few reductionistic strokes, circularity sets in, and a closer scrutiny reveals that there is no way to reduce the scientific edifice to some form of foundation.

The whole notion of foundation is suspect. What would it mean to posit a single grounding layer of explanation at the bottom of reality? Yes, it may well be that a single unified theory will be found that can be seen to underlie all the diverse types of particles and interactions that physics knows about at present. However, it is an altogether different question to ask how we can bootstrap ourselves up from such a "fundamental" understanding to a more "applied" knowledge of higher levels of organization, of every-day physics and chemistry, all the way to life, to intelligence, and to conscious experience.

Apart from the "gap" that opens wide between different layers of explanation (Anderson's "more is different"), there is an altogether different problem. Sure, within a physics view of the world, we can try to ground physical reality with a few simple rules and symmetries. But what does it mean to take that stance, to "go into" the world of physics? The whole notion of "doing physics" is a particular project. A very interesting project, for sure, and a very powerful project, as we have seen from its applications, for better and worse. But still, it is a project, one among many.

The project of "going into business" is another project. And a business man or woman naturally tends to view the world as one complex business project, just as surely as a physicist tends to see the world through the eyes of physics. An artist may see the world as one huge kaleidoscopic set of art frames, and so on, with an overwhelming choice of projects we can choose from. What is so special about the project of "doing physics"?

The traditional physics position is as follows. Physics gives a detailed description of the structure of objective reality. And the hope is that future improvements in our understanding of physics, and its derived disciplines layered on top of physics, would asymptotically lead to a complete understanding of objective reality. And here is the intended coup of the physics project taking over all other projects: physics tends to see "projects" as subjective tendencies that form part of the way that brains function, while brains are part of physical reality, and physical reality is what the project of physics studies. Voilà: the project behind the project of physics is to stage a palace revolution, namely to appropriate all other projects as ultimately derived from the project of physics.

I guess it is clear that I am somewhat skeptical of this move of appropriation. Ultimately the conflict here boils down to the question of where knowledge comes from. Is knowledge something that "emerges" out of complex physical systems called brains (in which case physics would have at least some chance to claim its foundational role)? Or is knowledge an inherent dimension of reality, more fundamental than any particular form of matter or energy, perhaps equally fundamental as space and time?

As I have tried to argue earlier, the latter view makes much more sense to me. But why argue? Does it really make any difference, in practice, how we view knowledge, and the status and source of knowledge? This question brings me to the more evocative aspects of a discussion of limits, and the related concepts of edges and boundaries.

IV. Limits, Edges, and Boundaries

Let us return to the question we started with. Is scientific knowledge in principle limited, either by the very structure of reality, or by the type of knowledge that qualifies as scientific? My guess for an answer would be "no." Of course, there is no way to prove that such an answer would be right or that it would be wrong. Instead, my strategy will be twofold: I will try to make my answer at least plausible, and I will also try to show that, as a working hypothesis, this answer has pragmatic value.

The No-Boundaries Working Hypothesis

Limits are the most paradoxical things we deal with in life. Whatever we encounter, whatever presents itself in the world or in our mind, including the ever-tentative distinctions we make between those two realms, any and all of that is given in terms of limits.

Limits are required for any type of structure, physical or verbal, real or imagined. Limits in the form of boundaries, separating one area from another, are what makes a structure a structure, rather than an amorphous something. In fact, we cannot deal with anything that is not structured in some way or other: notions such as "amorphous" or "something" are themselves set apart from other notions, and their meaning is thereby limited and, indeed, structured. Whatever presents itself in our experience is structured by dichotomies: inner and outer, blue and not-blue, good and not-good, and so on. The boundaries that separate them form the very texture of reality as we know it: without limits no knowledge.

The paradox of limits lies in the fact that limits combine two opposite functions: setting apart and joining. They divvy up the world of appearance (mental or physical, mathematical or verbal, or of any other type) into separate areas, in intricate and overlapping hierarchies. But at the same time they structure the interrelationships and communication channels between the pieces into which they have just carved up the world.

A cell wall protects the cell from its environment, and at the same time is a two-way input-output channel, connecting the cell with its environment and allowing metabolism, growth, and production of new cells through cell division. The finite number of rules in arithmetic, separating symbol manipulations into correct and incorrect ones, allow an infinite number of correct calculations to be carried out. Even in modern art, an area where boundaries have perhaps become least defined, each piece of art is given through some form of limit or description, at the very least through the way in which it is presented as set off from the rest of the Universe.

The paradox of limits can thus be summarized: we can view boundaries as bridges. This idea, that boundaries can function as bridges, can be fruitfully recognized and applied in all walks of life, in any type of analysis or activity. We already encountered the business notion that "every crisis is an opportunity," a very concrete example of the paradox of limits. In scientific research, too, anything unexpected, anything that seems to spoil the expectations of the experimenter working in the lab, or the theorist working with pen and paper or computer, is welcomed. Anything that seems to block further progress is a candidate messenger signaling potential progress into new, possibly totally new, areas.

Another way to interpret the paradox of limits is to say: there are no limits. This no-limit hypothesis denies the existence of any ultimate limits. Limits here are defined as enclosing something, setting part of the world apart from the rest, by making it off-limits. In a relative way, limits can and should succeed in doing so. But at the same time, any limit that is defined within one framework can be transcended within another, wider or simply different, type of framework.

Stating this in another way, I very much doubt the existence of any ultimate boundaries. This hypothesis cannot be proved or disproved, but we can work with it, using it as a guide in our considerations and actions. As a working hypothesis, it can encourage us, stimulating a creative attitude. Like artists, stepping over the bounds of the received way of dealing with things, we can see opportunities everywhere.

In science and technology, research "at the cutting edge" is what is valued most highly, the locus where important new results can be expected. At this cutting edge, researchers play the modern role of heroes in ancient myths, who placed themselves squarely on the edge of the unknown (with a few minor modifications, such as dragons having turned into tenure committees).

Exploring nature, through science, or exploring one's own nature, through philosophy or religion or arts, all these activities are a form of "pushing the envelope." We proceed towards the edge of what is known, and discover that an edge is not something to fall off of: borders are bridges, problems point to solutions. This is the marvelous key to the human condition: close and honest inspection of a border line sooner or later shows new openings in what we thought to be a tightly shut case.

Honest here means: authentic. Inauthentic scrutiny of a perceived border is characterized by a tacit acceptance of the border as border, a form of laziness and conspiracy with received wisdom. In contrast, authentic scrutiny regards a border as a valid boundary of past knowledge, something that may have had a useful role in the past, much as a cast for someone with a broken leg, or a play pen for a child growing up. Useful as these various devices may have been in the past, there comes a time to try to disregard them, in order to open up the received boundary, and go beyond.

This attitude is really a form of working hypothesis. Until new territory is found beyond the existing boundary, we cannot possibly "prove" that it is possible to cross the boundary, let alone dwell beyond. Until the current boundary has lost its boundary character, its non-boundary character is merely a hypothesis.

Another way to phrase the no-boundary working hypothesis is to say "we are free, we can never be pushed in a corner; there are no corners." In other words, there is always a way out. Perhaps not in the direction we thought of at first. Perhaps in completely unexpected directions. But whatever the situation, there are always new degrees of freedom to be found, new dimensions to be explored. And what looks like a hopelessly encircled situation within a limited number of dimensions can always be seen to be fully open in the direction of new and unsuspected orthogonal dimensions.

There are two main arguments that support the no-boundary working hypothesis, one based on logic and the other on history. The logical one is this: every boundary is a boundary within a certain setting; recognizing a boundary as a boundary already places the boundary in the realm of the known, making it an object of exploration. As soon as we can sit on the wall of the known, we can look out in the opposite direction of the known, and at least catch some glimpse of something that is there. After all, the unknown has at the very least one element in common with the known: the boundary. Here is the gist of the argument: acknowledgement and inspection can turn boundaries into bridges.

The second argument, based on history, runs as follows. In the past, how often have we not seen a firm belief in boundaries crumble! Man was not meant for this or that; nature was restricted to be such or such; this or that race or social class had its own limits. Which of these limits have been left in place? Those that have, upon closer inspection, are most often the result of a renaming of the terms involved. In physics, "energy" is still conserved, but this new concept of "energy" encompasses much more than heat and motion, it now includes matter as well. In politics, all "men" are equal now includes a concept "men" as "men and women" in a way that was nearly inconceivable a few hundred years ago.

Whether one is tempted by the evocative form of logic, or the lesson from history, or simply by a curiosity to try out this no-boundary working hypothesis, it is fun to test this hypothesis as critically as possible. The type of questioning that this hypothesis suggests is the most radical form I can think of: a form of inquiry that puts into doubt any and all barriers. Each barrier is seen instead as an invitation to freedom.

On Edge

Life at the edge has always had an appeal, in stories, fairy tales, or epics. Life at the edge of an enchanted forest, or at the edge of a vast desert or mountain range. Or life in a port, at the edge of a vast sea, stretching out to unknown shores; a sea fraught with peril and also beckoning, beckoning beyond the borders of the known.

Martial arts teach a practitioner to be on edge, in an open and attentive way. Expecting nothing in particular, fully open for whatever new situation presents itself, the known past has played out its role, and the unknown future is greeted with open arms. The practitioner is present, here in the present, at the edge between the known and the unknown.

In contrast, a more usual state of presence is to be fully ensconced in the past. Safely residing in the known (complete with insurance policies against assorted known but uncertain threats) the usual state of presence is not very present at all; it is a form of enduring past. Compared to living on the edge of experience, such a state with a cozy past-based perspective seems like a form of sleep, or even death. No wonder that all cultures have myths and parables that encourage us to "wake up"—to a life "on the edge."

Waking up is an image that occurs in all religions. Being born, or being born again is an even stronger image, familiar in our own Christian heritage: the New Testament's advice "let the dead bury the dead" refers to a life off-edge as "dead," as opposed to those who are truly alive, living on-edge. When Buddhism talks about life as a dream, it is off-edge living that is fully submerged in the dream, and it is on-edge where awakening can occur.

The notion of living "on edge" can take many forms of meaning, on a grand scale as well as on the level of everyday life. The expression "life at the frontier" offers one way of talking about life at the edge, with a grand ring to it. This frontier can take many shapes: an actual battle field, or newly explored territory, or newly conceived activities.

In one way, all of us lead a frontier life all the time, namely in the time dimension, where each moment of being alive is a moment at the frontier. In the very present, there are many years of our life history stretching back in time, and there are billions of years of our Earth's history. All our factual knowledge is concerned with the past. But in the other time direction, even the details of the next second of our life remain unknown—until they are transformed from future to past details.

In yet another way, we all live a frontier life. We always live at the edge of the unknown. Even in our own bodies, there are far more processes going on at any time than we are aware of. And in our immediate neighborhood, we always straddle the borderline between the known and the unknown. And with the pace of change having increased considerably in the past few centuries, both the known and the unknown have become accentuated more and more.

One expression of the borderline nature of modern life is provided by the rapid revolution in information technology. Soon each home will become an information port, an active nerve center for reception as well as generation of data of an as yet unforeseeable variety. This will be a step into the unknown, but we do have an interesting historical parallel, namely the change from travel by ship to air travel.

Only a hundred years ago, port cities played an important and unique role in the geography of a country. Only a seaport gave access to travel to those remote countries that could not be reached overland. Only from the rim of an island or continent could a journey start to other continents or islands.

The invention of the airplane has brought unexpected competition, and has effectively turned every city into a port city; the whole surface of the Earth has effectively been turned into coast line. Airports can be built everywhere on Earth. Even the most remote places have room for at least a helicopter to land, if not a small airplane. In a sense, the border of a continent has stretched itself out over the whole of the inland region: each place on Earth now finds itself "on edge."

The inventions of telephone, radio, and television have further brought access to far-away countries. Contact with overseas is now possible without either boat or airplane: each home has already become an information port, even though the exchange of information so far has still been rather primitive. But before long each of us will have the opportunity to get on the road, on the "information highway" with its growing planetary web of branching and connecting paths. It is the latter aspect that makes this "highway" so interesting, since it presents us with a system of alleys, byways, and local trails.

Perhaps a better metaphor is an "information sea," since a road suggests a far too narrow and one-dimensional picture of what is possible in information land. Indeed, a user of the "information sea" can quickly feel him or herself at sea, lost between the tidal waves of information that can be unleashed at the push of finger, with only a few key strokes.

With each place on Earth turning into a port, our life is being turned "on edge" in various ways. New ideas and new possibilities are being "ported" into our familiar surroundings. All our various landscapes are changing rapidly: job situations, friendships, voting procedures, ways of entertainment, and almost everything else will be affected by the information revolution in ways that we cannot possibly imagine.

Every technological revolution has changed society as well as individual lives in ways that were never anticipated. And in those cases that predictions were made, these predictions almost invariably turned out to be completely off the mark. To mention one glaring case: remember 1984, the year of Big Brother who would keep electronic surveillance over all of us?

That scary vision was understandable in the late forties, when electronics components were so expensive that only large governments could seem to afford them. But now, with everyone getting access to computers, what is the case? Exactly the opposite! Now citizens can band together without any possible way of control by the various governments. Now governments are scared, and scratch their heads in vain in search for solutions to keep their citizenry down on the farm, away from the information sea—but in vain.

In 1989 there were not yet enough fax machines in China to convey widespread information about the repression at T'ien-An-Men square, even though Chinese from Hong Kong managed to reach thousands of friends and business relations this way. Now there is no such limitation any more in China, and soon no longer in any country on Earth.

No, life in a port city is not a bowl of cherries for would-be dictators. My guess is that their days are numbered, and that a constructive and balanced form of anarchy will finally be possible, for the first time in history—not only possible, but necessary: "once they've emailed to Paris, how are we going to keep them down on the farm" must be the sigh of many a politician. All politicians will become civil servants, they have to, or else: information, and hence termination of their privileged position.

Plurality

Returning to the no-boundaries working hypothesis, we have to face the question of what it means for scientific knowledge to be not limited, in principle. Does this imply a vision of the future in which science will cover everything, as a steamroller flattening other fields of knowledge, subordinating them to a given fixed "scientific method"? Obviously, such a notion would be far too narrow. Science is something dynamic, something evolving, something that escapes any attempt to catch it into timeless dogma.

Scientific research is a style, constantly shifting, opportunistic, one that in principle can be adapted to study anything. Modern physics is a perfect example of how a form of rationality can be tailor made to fit something that at first may have seemed very strange and rather unreasonable. Even something so counter-intuitive as quantum mechanical phenomena, where one particle can be thought to "exist" at different places at the same time (not to mention being able to go back in time), has been incorporated in the formalism of physics.

It is an interesting thought experiment, to imagine a physicist from the eighteenth or nineteenth century visiting us in the twentieth century. A fair fraction of the fundamental concepts that such a person grew up with have been dethroned. Absolute space and time are out the door. Energy conservation does not hold strictly true anymore for sufficiently short time intervals. Deterministic systems are known to show truly chaotic behavior. We could continue the list for quite a while.

But what is most interesting is this: such a visitor, once made familiar with our current ideas, would probably have little difficulty recognizing our enterprise as still being scientific. A good scientist recognizes a reasonable scientific approach, independently of the details, even if those details have earlier seemed to be absolutely essential.

Science thrives in an atmosphere of open rationality. This openness is something that allows responses in a playful way to any changes that may be required. Rationality implies an adherence to a form of logic, a rational framework that seems most apt for the situation at hand. Not cast in stone, such a logic can and will continuously change in major and minor ways, as long as scientific research is alive and active. Not tied to any *a priori* assumptions, an open rationality is the ideal vehicle for science.

It is interesting, and not a little ironic, that many of the so-called "softer" sciences, such as psychology and sociology, have tried to borrow what they considered to be hard-nosed physics type approaches to research—only to find that many of those techniques had either been superseded in physics itself, when new data required new forms of explanation, or were actually never used as advertised.

What happened earlier this century, was that many sociologists, psychologists, economists, and others tried to incorporate elements of the "method of physics" into their own research. The problem here was that many social scientists really believed the polished positivistic stories about the alleged way physicists work, in their systematic way of defining first principles, strict rules, and neatly organized step-by-step approaches to as-yet unsolved problem areas. They were not alone in being fooled. In fact, most beginning students in physics are fooled as well by text book accounts of how progress has been made over the last few centuries.

To some extent, it is perhaps unavoidable to fool students this way, at the start of their training. It would be rather impractical if one had to go through the details of the few hits and many misses of the historic process of building up physics, through the politicking and personal rivalries, the many emotional reactions that either helped or hindered progress at certain times, etc. And also, it is in practice not very harmful, as long as the teacher presents the text book material as what it is: encyclopedic in nature, and not at all a fair account of how this knowledge has been arrived at.

In any case, when physics students begin to do their own research, the message is quickly driven home to them that actual research is far messier and much less systematic a process of induction and deduction, than the text books had led them to believe. The real victims of the myth of "the scientific method" are therefore not the students inside physics, but researchers in completely different fields, wanting to actually apply this mythical animal.

The moral of the story is that each field and subfield needs its own formalism. And while there sometimes are unsuspected parallels between seemingly unrelated fields, such surprises are more the exception than the rule. What is needed is a form of pluralism, an approach to science that lets a thousand flowers bloom, each in its own way.

Role Playing

A pluralistic world view does not automatically exclude any and all forms of unity. To talk about plurality and about different forms of logic presupposes the possibility of comparing them, at least to some extent. Some form of middle ground has to be present, as the condition of possibility for such a discourse. No matter how incompatible the logics and fields of investigation may seem, some form of connection, however tenuous, must be present.

At the same time, it is important to be aware of the trap of positing facile universal features, ironing out differences that seem less important. Such "flattening" of "uninteresting" details almost always stems from the use of a particular political agenda, often tacit, and perhaps completely unconsciously present, handed down as received wisdom through education. Whatever the origins are of particular attempts to cut the world down to size, in order to fit within a pregiven unified view, such attempts almost always imply forms of violence.

Unity thus implies responsibility. We cannot avoid talking about unity, since it would be utterly naive to try to treat any single event completely on its own, as if nothing connected with anything else. We have to choose dots to connect, and we have to acknowledge patterns we see within the plurality that we encounter in our life. And any form of connecting carries with it a responsibility for the consequences of the connecting activity. This, then, brings us to the question of ethics.

Science that does not have any ethical implication can be useful, but cannot claim in any way to describe all of reality, since clearly some form of ethics is part of our world of experience. On the other hand, ethics as a set of arbitrary commands, either ascribed to a superhuman source or to a biologically useful set of general rules, is not satisfactory either.

Spinoza may have come up with the best suggestion so far: that a deep understanding of reality automatically gives a deep understanding of ethics. His main work was entitled "Ethica." And, significantly, he came up with an organic view of the world, a form of panpsychism if you want, very far removed in spirit from his main source of inspiration: Descartes. Spinoza is much closer in spirit to our modern sense of ecology than Descartes, the vivisectionist who regarded even mammals as mere mechanisms, as clockworks that could be freely taken apart in order to study their mechanism.

The particular view Spinoza came up with may not be too attractive, perhaps, and can strike us in some sense as outdated. His lack of emphasis on the historical dimension of life, for example, is something that does not appeal to most of us anymore. However, his open-mindedness is still remarkable: avoiding the dualism of Descartes, he combined a non-dualist monism of substance with a form of infinitism of attributes.

The notion of ethics seems to be tied up with the notion of role playing. A play can have both the connotation of something serious as well as something frivolous. In order to play a good play, one has to be quite serious. But taking a play or a game too seriously may not necessarily be the best strategy. A somewhat more detached view may actually a better player make.

Starting from a no-boundaries hypothesis, and seeing everything as a play, as an interplay of forms of software, everything is questionable. There is no hardware and there are no hard limits. All experience, in principle, is equally "valid", and equally material for a radical form of questioning reality. Poetry is neither more or less "real" than physics, for example. Feelings and reason are equiprimordial [11].

What is this world we are living in, and who are we? In order to come to terms with such questions, we can switch from an inquiry as to "what" to a more revealing inquiry as to "how." How does this whole world arise in the way it does, in the way it appears to us? And how does our notion of who we are arise in that same experience in which the world appears as well?

Asking such questions, we can find a tentative answer to what it means to say that something "is." There "is" a cup, there "is" joy, there "is" form and function and value. Whatever appears, it has to make some form of sense to us, in order to qualify as something that "is." Even utter chaos or non-sense presents a form of sense (namely: chaos, nonsense). So, for us, "what is" is the direct result of identifications we have made.

In a very real sense, the world we find ourselves in as well as what we believe ourselves to be are the result of ideologies, of identifications that are, in an ultimate sense, highly questionable. This is not to say that there is not a practical value to our usual interpretations. Of course there is. Of course we need a large amount of knowledge about the world in order to be able to function in it. It is only when we forget the role-play character of all that we consider to be "real" that we get into trouble. Having labeled something as "real," we make it more difficult to delve deeper, below the roles that have been labeled as "real." The concept of reality is like a terminal station, whereas the notion of role playing beckons to continue the ride to the next station stop, and the next, and the next ..., in a dynamic unfolding that sees no need to freeze and tie down what appears to a flat level of a single, unique, and absolute reality [12].

As long as we avoid the habitual identification with the roles that are being played, the answer to the question of "what is" retains its multi-layered character. At each moment, the question of "what is" can be answered from within the play in which something takes significance, as well as from within a larger play that embraces the framework of the more specialized "play within the play," or an even larger play, and so on.

For example, a pawn within a game of chess has to obey certain strict rules and consequently is caught in a situation with severe limitations. But when seen as just a piece of wood, it literally can be moved anywhere on the board at any time, or it can even be moved off the board altogether. And as a piece of wood, it can be again be viewed in many ways. If it were carved in an unusually intricate way, it could be deemed worthy to be exhibited as a piece of art, even though its role within the chess game would not be changed at all. Or it could be seen as "just a piece of wood" and perhaps thrown in the fire as a consequence of being denigrated to a piece of firewood. Clearly, the question of "what is" is highly contextual.

Freedom from Identification

Earlier I mentioned that each barrier can be seen as an invitation to freedom. Freedom here means this: a freedom from identification. When all barriers are seen to be relative, contingent upon the framework we choose to adopt, the barrier-nature of barriers is recognized as resulting from identification with the corresponding framework. Letting go of this identification does not mean a destruction of the barriers, rather it means a breaking free from previously unquestioned identification. Recognizing a movie to be "unreal" does not imply that we have to walk out of the movie theater. On the contrary, it is through a recognition of a play as a play that we can fully appreciate the drama being played out.

In a nutshell, this is my motivation: using the no-boundary working hypothesis as a starting point, my goal is to explore the possibility of a freedom from identification.

Here identification is any form of barrier building in which the barriers are considered to be absolute. Other terms can be used, instead of "barrier," depending on the circumstances, such as edge, or border, limit, frontier, horizon, boundary, wall, or skin. In all cases, however, a barrier does have a relative function. Just as a cell wall or a human skin has both a protective and a communicative function, each boundary can be explored in these two modes.

Boundaries can turn into bridges, the moment we acknowledge the context dependency of the limiting role that boundaries play. And as I argued before, I see role playing as that which allows boundaries to be transcended.

Freedom from identification is the immediate result of seeing through the propaganda attached to the role playing, the propaganda that suggests that the roles are "real," more than relative to their contextual situations. Freed from the massiveness of a given outside reality, ethical questions of what "ought to be" can then be seen in a new light. In a very practical way, questions of change can be dealt with in a fluid way. From a contextual viewpoint, we can be responsive to the situation at hand, without the need to recite ideological or religious scriptures or other codified systems of problem solving.

At any time, we can view anything in its "being" aspect, as the role that is being played, as that "what it is." But we can equally well view it in its "non-being" aspect, in its aspect of openness or emptiness. From the point of view of the play, the player underneath the role being played is simply not there. In a drama, there "is" a king. The actor "is not" within the rules of the play. Within the play the actor steps aside, disappears, to let the king show through. But when we step outside the play, the king has vanished, has completely lost his base, his foundation of existence. We then see that, at bottom, the "king" has been an empty notion all along. something being played but not ultimately "real" in any sense.

It would seem that, instead, the actor is the real person, rather than the king. But what if the actor unexpectedly gets fired, soon after the play? Then the next layer drops. The actor disappears as well, and a jobless person appears instead. Is there a core that remains unchanged? Flesh and bones, or molecules, or a life history constructed as a vast web of connected past events and relationships between events? Or are any and all of these in turn the results of further attempts at role playing and play interpretations? I suspect the latter. Let me try to sketch what that may mean.

Stated in the most radical way, each subject or object, human or physical object or abstract idea or whatever, is playing a role. And what we identify as playing the role is itself playing a role. We are part of a great drama of role playing, with roles within roles within roles—without anybody or anything "home" underneath; without any stable and final foundation to bolt things down upon.

This emptiness is what allows anything to appear in the first place. The notion of emptiness is truly the most positive notion we can come up with, the one notion that is least notion-like, if we can resist the temptation to conceptualize it [13]. Only emptiness can provide full openness. And this openness is fully accessible as soon as we look through the layers of role playing that tend to obscure the underlying openness.

The funny thing is, though, that the obscuration has never happened in the first place. Within a play a king can be a powerful person, but once we look from a vantage point outside the play, what is left of the power of the king? Even if we would try to strip the king of his power, to rebel against him in order to overthrow him, we would not find any handle. There would be nothing to fight against. Emptiness and openness do not offer a place for a sword to cut into.

Freedom from identification is something extremely paradoxical. Each time we gain an extra measure of authentic freedom, we realize that we have been free all along, that we have not found anything new at all. Take the example of a moth flying around a lamp. Physically, the insect is completely free to fly away, any moment it wants. But the problem is, it doesn't "want" to. And while we may consider moths to be programmed biologically, we cannot maintain the same excuse for ourselves. From within the play (of being obsessed with flying in circles), there is no freedom. From outside the play, there never have been any prison walls.

And this is not just a fancy form of wishful thinking. It applies to anything in daily life. Once we wake up to the tentativeness of the world, and to the contingency of being, the massiveness of the world can drop away, gradually or suddenly. A lightness of Being can make itself felt, even in the marrow of our bones, as attested by so many pieces of world literature; not to mention many of the great mythologies of various cultures. All this may sound mystic (a label and verdict that is very suspect in scientific circles). Be that as it may. If an appreciation for appearance as such, before any form of interpretation or reduction, is classified as a form of mysticism, then my view should indeed be labeled as mystic. But I rather prefer to label it as a form of empirical philosophy, or more accurately: radical empiricism [14].

There are interesting parallels with mysticism, though. If a scientist looks at the claims of a traditional religion, he or she is likely to be rather skeptical at the seemingly arbitrary boundaries that are acknowledged by the true believers of that religion. To take one example, the believer may claim that a particular temple ground is "holy," and that there is a clear distinction in sacredness between what lies within the perimeter of the temple area and what lies outside. The scientist may object that the molecules inside the temple are the same as those outside, and that no scientific analysis is likely to yield any measurable distinction that could justify the presence of a definitive limit, separating the sacred from the non-sacred.

Such a scientific attitude would be very reductionist, and would probably not convince the believer. The scientist in turn might well be ready to concede a contextual value to the notion of sacredness. But it is interesting to take up the reductionist conclusion, and turn the tables. The first, naive, interpretation of the absence of a distinction between temple and non-temple would suggest that nothing is sacred. But this is only one way to react to the dropping away of a limit. There is an alternative: we may equally well conclude that everything is sacred. Rather than limiting appreciation to a particular spot, we can follow the examples of mystics of all ages, who have never tired telling us that there is no such thing as ordinary, finite, non-sacred things and events. Anything can be viewed in its proper aspect, as an open gateway to a boundless reality. It is here that science and mysticism meet, in an outlook that is literally limitless.

Acknowledgments. This work was supported in part by a grant from the Alfred P. Sloan Foundation to Piet Hut and Roger Shepard, for research on limits to scientific knowledge. I thank Ron Bruzina for many discussions on the topic of foundations and horizons. I also thank Roger Shepard, Bob Rosen, Otto Rössler, and Mel Cohen for many discussions on the topic of the paradox of limits. I thank Arthur Egendorf, Thomas Herbig, and Eiko Ikegami for comments on the manuscript of this chapter.

Notes and References

[1] A note about philosophy here: granting knowledge the same ontological status as space and time, and a more fundamental status than matter, may label me as an idealist, as opposed to a realist. However, I am not happy with these categories. If anything, I would prefer to classify myself as an radical empiricist, along the lines of William James's *Essays in Radical Empiricism* (1912), reprinted in *Essays in Radical Empiricism & A Pluralistic Universe*, by W. James [1967, Peter Smith]. Within twentieth-century Western philosophy, I feel most attracted to Husserl's ideas (cf. Husserl, E. 1913, *Ideas for a Pure Phenomenology*), but I am also sympathetic to many non-European systems of thought that approach knowledge in a less derived way than we normally do (cf. Nishida, K., 1911, *An Inquiry into the Good* [1990, Yale Univ. Pr.]). Specifically, the notion of viewing knowledge on a par with space and time has been suggested by Tarthang Tulku, 1977, *Time, Space, and Knowledge* [Berkeley: Dharma Publ.]

[2] For a vivid description, refreshingly unsophisticated and down-to-earth, see Harding, Douglas E. 1961, 1988 *On Having No Head* [London: Arkana].

[3] Husserl, E., *The Crisis of European Sciences,* 1970, Northwestern Univ. Pr., p. 137.

[4] Harvey, C.W., *Husserl's Phenomenology and the Foundations of Natural Science,* 1989 [Athens: Ohio University Press], p. 233.

[5] Nishida, K. *op. cit.,* p. xxx.

[6] For a view from across both sides of the fence, by someone who has received full training in both Eastern and Western philosophy, see Mohanty, J. N., 1992: *Reason and Tradition in Indian Thought* [Oxford: Clarendon Press], pp. 282–299.

[7] Romanyshyn, R. D. 1989, *Technology as Symptom and Dream* [New York: Routledge].

[8] Anderson, P.W. 1972, *Science,* **177**, 393.

[9] Polanyi, M., 1962, *Personal Knowledge* [Univ. of Chicago Pr.], p. 382.

[10] Rota, G.-C. 1989, *The Monist,* **72**, 70.

[11] cf. Rota, G.-C., *op. cit.*

[12] There are various approaches in robotics that reflect this active type of reality construction; for a recent introduction and overview, cf. Franklin, S. 1995, *Artificial Minds,* [M.I.T. Press].

[13] Among the recent explosion of books on emptiness, one place to start would be *Emptiness of Emptiness,* 1989, by C.W. Huntington [Honolulu: Univ. of Hawaii Pr.].

[14] This term was introduced by James, W. 1912, *op. cit.*

Chapter 7

COMPLEXITY AND EPISTEMOLOGY

HAROLD J. MOROWITZ

I. The Classical Period

The period of Galilei and Newton (1600–1700) marked the beginnings of modern science. The new method of approach that was instituted consisted of moving from observations to theories to the prediction of additional observations. The agreement between the predictions and the second set of observations was the ultimate validation or invalidation of a theory. The theories of physics in the early period largely dealt with mechanics and came to be formulated in the language of algebra and differential calculus. Using the syntax of mathematics, physics took on a Platonic mode that was enormously successful in fields such as celestial mechanics and carried over into mechanical engineering and the industrial revolution.

The rise of deductive science in the time of Galilei overlapped a period of the flourishing of inductive science or the natural history stage of inquiry. This mode of thought was set forth in the *Novum Organon* by Francis Bacon. His method stresses observation, description and classification as the principle methodology of science. Written at the beginning of the 17th century, Bacon's book sets forth the standard inductive protocol for obtaining knowledge. Although Galilei's approach took over in physical science, Baconian induction persists in areas of the natural and social sciences and, indeed, in 1934 Karl Popper began his critique of the philosophy of science with a discussion of "The Problem of Induction."

This enormous success of Newtonian physics and its extension to electricity and magnetism led to attempts in many other disciplines to try the same approach. Chemistry was the most successful leading to formulation of the atomic hypothesis and eventually to establishing most of the phenomena of physical chemistry in terms of thermodynamics and quantum mechanics. A period of neoplatonism followed the success of Newton's theory in which the assumption was made that the study of all disciplines could be carried out using the same mode of reasoning used in classical physics. The book *Mathematical Biophysics* (1938) by Nicholas Rashevsky was an attempt to formulate many problems of biology in terms of second-order differential equations subject to boundary conditions. Rashevesky's work had little influence on the science of biology that was beginning the molecular revolution. Social scientists also tried to mathematize their disciplines developing numerical metrics which often seemed to have a somewhat ad hoc character.

II. Epistemology

The epistemological character of post-Newtonian scientific investigation went through a long period of debate, often between the positivists and the naive realists although many other voices were heard. In the mid-20th century two statements of the philosophy of physics became central to understanding theory. Although they are closely related, one is prescriptive and the other is descriptive. Central figures in these approaches were Karl Popper and Henry Margenau. Our discussion of these preeminent philosophers of science will very briefly state their views

The first edition of Karl Popper's book was published in German in 1934. The English edition, *The Logic of Scientific Discovery,* appeared in 1959. The initial version came at a time following a 30-year upheaval of the ideas of classical physics following the formulations of quantum mechanics and relativity.

Popper starts by pointing out the limits of induction and says that science proceeds from an analysis of a problem to a theory, a creative intuition, to deductions which must be checked against experiments. If the theoretical deductions are falsified by experiment the theory is falsified and one starts again. Theories that are not falsified are temporarily the content of empirical science. He states:

> It should be noticed that a positive decision can only temporarily support the theory, for subsequent negative decisions may always overthrow it. So long as a theory withstands detailed and severe tests and is not superseded by another theory in the course of scientific progress, we may say that it has 'proved its mettle' or that it is 'corroborated' by past experience.

He further elaborates this view:

> But I shall certainly admit a system as empirical or scientific only if it is capable of being tested by experience. These considerations suggest that not the verifiability but the falsifiability of a system is to be taken as a criterion of demarcation. In other words: I shall not require of a scientific system that it shall be capable of being singled out, once and for all, in a positive sense; but I shall require that its logical form shall be such that it can be singled out, by means of empirical tests, in a negative sense: it must be possible for an empirical scientific system to be refuted by experience.
>
> My proposal is based upon an asymmetry between verifiability and falsifiability; an asymmetry which results from the logical form of universal statements. For these are never derivable from singular statements, but can be contradicted by singular statements. Consequently it is possible by means of purely deductive inferences (with the help of the modus tollens of classical logic) to argue from the truth of singular statements to the falsity of universal statements. Such an argument to the falsity of universal statement is the only strictly deductive kind of inference that proceeds, as it were, in the 'inductive direction'; that is, from singular to universal statements.

Popper's essentially prescriptive view of scientific validation has achieved wide acceptance, particularly in the physical sciences. It has also been influential in related disciplines.

III. Physical Reality

The Nature of Physical Reality by Henry Margenau approaches the same problem as Popper but in a descriptive way. Margenau was a distinguished theoretical physicist during the period of the growth of quantum mechanics. He examined what physicists had been doing for the last hundred years and asked what philosophical assumptions are necessary to account for the success of physics in explaining aspects of the natural world.

The outline of his epistemology has a formal similarity to Popper's in moving from observation to intuition to theory to observation. As with the earlier scheme, falsification negates a theory. Margenau is much more concerned than Popper with theory structure and the question of how to choose between competing theories that cannot be distinguished by falsification. Figures 1 and 2 illustrate the two views. Margenau recognizes many levels of abstractions from objects to microscopic entities to particles to force fields to probability distribution functions and other abstractions. All of these are regarded as theoretical constructs. Next are introduced the metaphysical requirements on these constructs. Again, we must note the descriptive character of these abstract ideas. These are

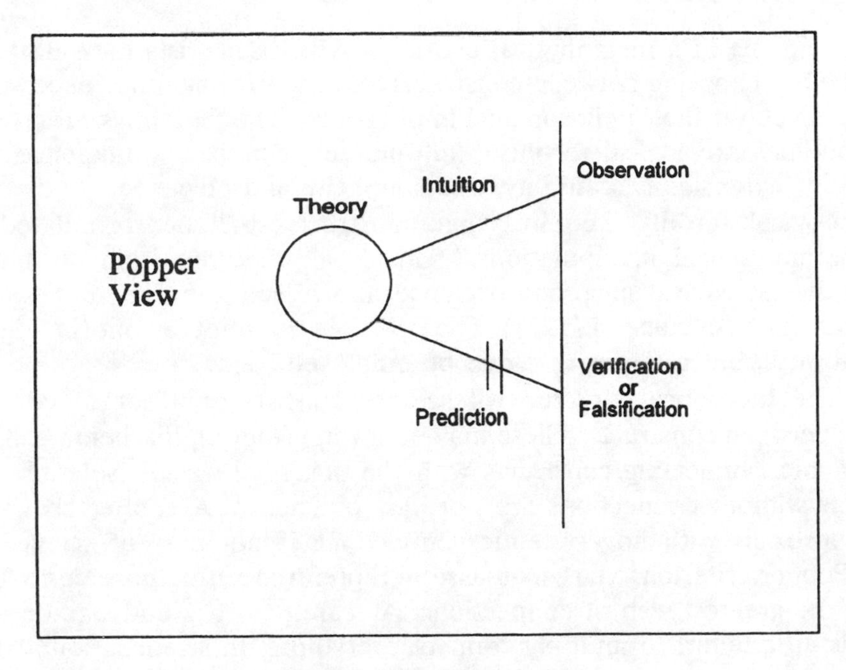

Figure 1. The Popper view.

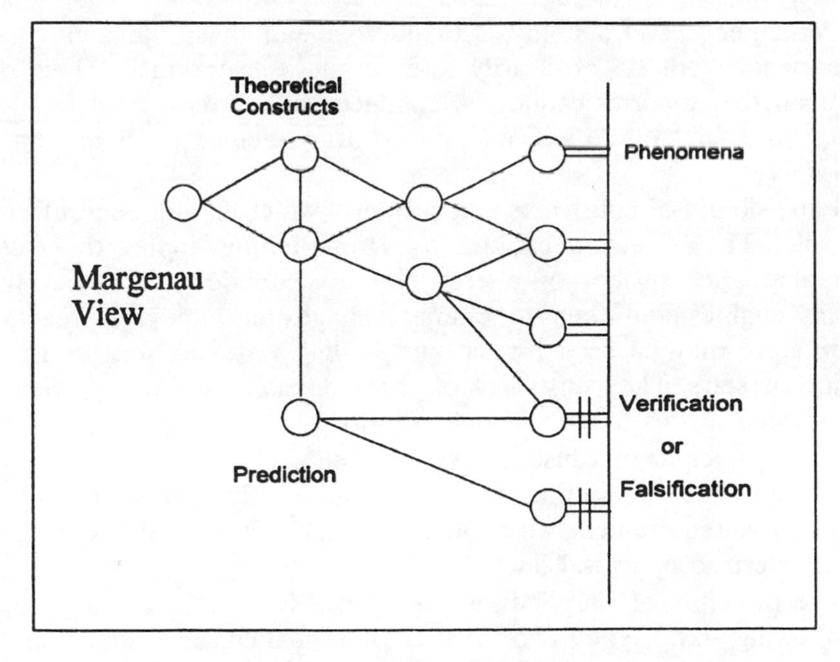

Figure 2. The Margenau view.

requirements of a metaphysical character which scientists have de facto applied in choosing between constructs and theories, without necessarily thinking about their philosophical implications. The metaphysical criteria he outlines are logical fertility, multiple connections, permanence and stability, extensibility, causality, and simplicity and elegance.

Logical fertility "requires that constructs shall be formulated as to permit logical manipulations." This is a weak requirement, but it separates science from pseudo science and rejects the "Anything goes" approach of Feyerabend (1975). The principle does not lay out the logical framework, but requires that one be employed.

The theoretical structure of science consists of a set of connections between constructs. These may be formal (logical, mathematical) or epistemic, connecting constructs with the observed world. Isolated constructs without connections are not part of science. A connected group of constructs with no epistemic connections is not part of science by the Popper criterion. Margenau asserts a preference for those constructs with the greatest web of connections. A search for a grand unified theory is an attempt to multiply connect everything. In science, confidence grows as constructs are multiply connected. We will return to multiple connectedness in discussing Perrins's work on the atomic hypothesis.

Permanence and stability is taken to mean that "the elements of a theory may not be arbitrarily altered to fit experience." This does not mean that theories cannot be changed; rather they must be stable throughout their application, without arbitrary change to fit this or that experience.

Extensibility of constructs is a property which allows generalization of results. Thus, Newton could move from falling apples to celestial mechanics. The concept of energy can be extended from mechanics to heat engines and then to energy changes in biological reactions. Temperature may be used for ice and boiling water as well as for the interior of stars. The constructs of thermodynamics seem particularly well adapted to this feature of extensibility.

"The principle of causality asserts that a given state is invariably followed in time by another specifiable state." By making causality a metaphysical requirement we choose constructs so that the system will be characterized by causal laws.

The principle of simplicity and elegance is stated by Margenau in the following way, "When two theories present themselves as competent explanations of a given complex of sensory experience, science decides in favor of the simpler one." This is, of course, an extension of Ockham's razor. There is also an esthetic element that is more difficult to define, because it is in the eye of the beholder. Nevertheless, scientists often recognize a theory as beautiful.

One persuasive feature of the Popper-Margenau view as practiced in physics is the precision of the verification as well as the high degree of connectivity of the constructs that have found their way into modern science. Two examples will serve to illustrate the above considerations.

The visible spectrum of atomic hydrogen yields a series of spectral lines of wave numbers

$$\bar{\nu} = R\left(\frac{1}{4} - \frac{1}{m^2}\right), \quad m = 3, 4, \ldots,$$

where $\bar{\nu}$ is the wave number, and R, the Rydberg constant, is experimentally found to be $109,677.58\,\text{cm}^{-1}$.

The theory introduced by Niels Bohr in 1913 predicted that

$$R = \frac{2\pi^2 me^4}{ch^3},$$

where m is the mass of the electron, c is the velocity of light, h is Planck's constant, and e is the charge on the electron. Putting in the best values of all the parameters, the theoretical R value agrees with experiment to one part in 10^4. The values of the four quantities come from many fields of physics and here show the high connectivity and the precision of agreement of predictions and experiment. This combination of connectivity and precision gave weight to the acceptance of a theory which had not been falsified.

A second example was shown in Jean Perrin's establishment of atomicity in his 1913 book *Les Atomes.* He presented sixteen different determinations of Avogadro's number, which all agreed within experimental error. These come from a wide array of physical techniques and converge on a common value. Again, falsification could have come at any of a large number of places and the coherence lends a strong support to the theory.

Table 1, adapted from Perrin's book, gives the sixteen independent determinations of Avogadro's number. In this table, the construct by Avogadro is connected: to viscosity by kinetic theory, to equilibrium thermodynamics by the Boltzmann distribution, to Brownian motion by Einstein's theory, to statistical mechanics by fluctuation theory, to

Table 1. Different determinations of Avogadro's number.

Observed Phenomena		Avogadro's Number
viscosity of gases (kinetic theory)		6.2×10^{23}
vertical distribution in dilute emulsions		6.8×10^{23}
vertical distribution in concentrated emulsions		6.0×10^{23}
	displacements	6.4×10^{23}
Brownian movement	rotations	6.5×10^{23}
	diffusion	6.9×10^{23}
density fluctuation in concentrated emulsions		6.0×10^{23}
critical opalescence		6.0×10^{23}
blueness of the sky		6.5×10^{23}
diffusion of lights in argon		6.9×10^{23}
black-body spectrum		6.1×10^{23}
charge as microscopic particles		6.1×10^{23}
	projected charges	6.2×10^{23}
radioactivity	helium produced	6.6×10^{23}
	radium lost	6.4×10^{23}
	energy radiated	6.0×10^{23}

electrochemistry by the Faraday and charge on the electron, to radioactivity theory by a series of detailed quantitative measurements and to black body radiation by Planck's theory. The convergence of values is a supreme example of connectivity in classical physics.

The persuasiveness of much of classical physics and chemistry stems from this kind of constant testing of theory by observation. In general the classical theories are computationally tractable and we ask for precision in verification. Of course, this kind of science tended to be confined to domains limited by analytical and computational techniques and tended to exclude biology and the social sciences.

IV. Complexity

The development of high speed computers during World War II and the postwar era and the particular rapid expansion of the application of computational techniques in recent years has led to a new kind of theory which does not fit into the philosophical mold of the Popper and

Margenau approaches and may require serious epistemological analysis before we know how to evaluate the results of a number of ongoing studies and assess their meaning.

Before getting into the computational domain consider some traditional fields of science where the falsification criterion is less persuasive: historical geology and evolution are examples. Again one starts with observations such as fossils, stratigraphic data, and properties of contemporary organisms. Theories are then constructed to account for the observations. These theories are only weakly predictive in regard to other observations and the predictions are at best qualitative. Evolutionary biologists are acutely aware of this situation and jokingly refer to their theories, or at least their colleagues' theories, as "just so" stories.

A school of rhetoric has stressed this storytelling nature of science. In the words of Donald N. McCloskey, one of the leaders of this school:

> Most sciences do storytelling and model building. At one end of the gamut sits Newtonian physics—the Principia (1687) is essentially geometric rather than narrative. Charles Darwin's biology in *The Origin of Species* (1859), in contrast, is almost entirely historical and devoid of mathematical models. Nevertheless, most scientists, and economists among them, hate to admit to something so childish-sounding as telling stories. They want to emulate Newton's elegance rather than Darwin's complexity.

What McCloskey ignores is that most scientists want to emulate the Newtonian approach because it follows the epistemological principles of the Popper-Margenau approaches and therefore allows it to go far beyond storytelling by precise numerical checks.

At the poles there are two kind of science: that which is subject to falsification by precise numerical comparison and that which operates in a much more qualitative fashion. The difficulty in the latter domain is the problem of deciding between theories all of which have some sort of qualitative agreement with observation. One could end up with a dozen plausible theories and no way of distinguishing among them assuming that they all pass the metaphysical criteria.

For domains such as evolution, ecology, economics and certain branches of physics that do not fulfill the criteria in a full quantitative sense a new computer driven mode of approach has grown up within the domain of complexity theory. The approach is still from observation to intuitive theory to observation to potential falsification. A new series of steps is introduced. Theory working with simple elements generates such a large array of computer generated outputs, that a second intuitive step is required: pruning or selection of algorithms that operate on the output of the theory. This is one of the core elements of the series of studies designated as complexity–the combinatorics of a set of simple constructs operating jointly produces an array of possibilities too large to deal with, sometimes transcomputational. To handle this enormity of outputs algorithmic pruning or selection is used. Genetic algorithms are a very good example of this kind of procedure. Fitness criteria are new intuitive theoretical elements that must be introduced. Under this mode of operation the successful theories may generate emergent properties not directly derivable from or perhaps even imagined from the simple constructs but by running the program. This emergence is one of the desirable features of complexity theory.

The epistemological loop in the complexity approach looks deceptively like the Popper-Margenau approach (see Figure 3) but varies in two very significant ways. One we have already noted is the introduction of a new step in the loop which has a different and not yet fully elaborated epistemological character from those previously discussed. The second is that the final outputs lack the precise numerical matching with precisely determined experimental quantities such as we demonstrated with the Rydberg constant. Indeed they are often vague and qualitative and come closer to McCloskey's view of telling stories. What we lack are criteria for distinguishing between conceptually different theories all of which may lead to plausible qualitative results for the study of a given phenomenon. One has a feeling that the theories that persist are those propounded by the best story tellers, and this leaves us a little uncomfortable.

This is not as negative as it may seem at first glance. Emergence is a powerful feature and might indeed be added to the metaphysical criteria in evaluating a complexity theory. It is an important feature and logical development of extensibility. The role of emergence can be seen in the following comments about how science has proceeded in disciplines such as biology that have a hierarchical character. Reductionism, the most favored approach, has been to seek the explanation of each hierarchical level at the next lower level. The clearest example is biology where the activity of tissues is explained in terms of cells, cells are explained in terms of organelles, organelles are explained in terms of macromolecules, and so forth. Beyond reductionism a more synthetic approach has long been a theoretical goal: the ability to predict the emergence of the behavior of a level from the rules governing the behavior of elements at lower hierarchical levels. Prior to the complexity mode of reasoning

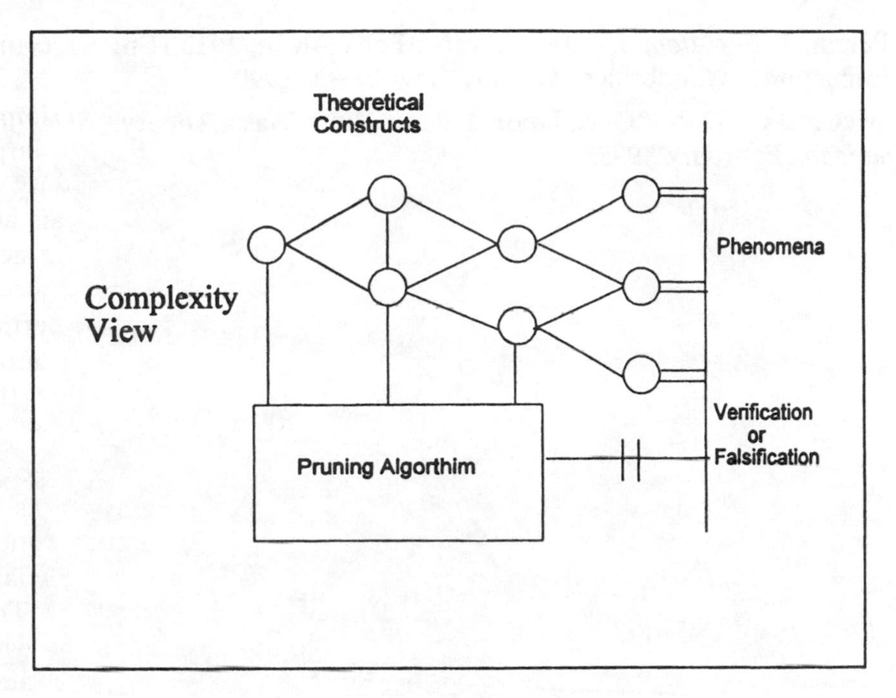

Figure 3. The complexity view.

this effort has been blocked at the stage of generating families of nonlinear equations that are analytically intractable. The ability to deal with emergence is thus a major step forward and deserves the utmost consideration.

The price we have to pay for the leap forward is finding ourselves in an unfamiliar epistemological domain. I do not have the theory of knowledge for the new approach to science, but the first step is to recognize the existence of the problem.

References

1. Rashevsky, N. *Mathematical Biophysics.* Chicago: University of Chicago Press, 1938.

2. Popper, K. *The Logic of Scientific Discovery.* New York: Harper and Row, 1959.

3. Margenau, H. *The Nature of Physical Reality.* Woodbridge, CT: Ox Bow Press (Reprint), 1977.

4. Feyerabend, P. *Against Method.* London: Verso Editions, 1975.

5. Perrin, J. *Les Atomes.* Paris: Librarie Felix Alcan, 1913 (English translation: *Atoms.* Woodbridge, CT: Ox Bow Press, 1990.)

6 . McCloskey, D. N. "Once Upon a Time There Was a Theory." *Scientific American,* February 1995.

Chapter 8

ON THE LIMITATIONS OF SCIENTIFIC KNOWLEDGE

ROBERT ROSEN

There are no impregnable fortresses. There are only fortresses that are badly attacked – Les Liaisons Dangereuses

I. Introduction

I do not consider myself a philosopher. I am a biologist, attempting to grapple with the Schrödinger question, "What is Life?" It turns out that this is not an empirical question, to be resolved through observation in a laboratory. Hence I am a theoretical biologist, not a "practical" one. Many of my experimentalist colleagues, who tacitly feel that science is consubstantial with laboratory practice, accordingly reject the question itself as unscientific, and it does not bother them that they cannot answer it.

I believe, on the other hand, that the problem is not only perfectly scientific, but that it is central. I also believe that it is in some sense solvable. Thus, I have developed a different view of what "science" is and what it is about than most of my colleagues. I have had to spend much time in exploring the capabilities and the limitations of many methods which have been proposed to give an answer to the question, but which do not in fact do so. These methods all proceed by replacing the "real world" by one or another artificially circumscribed one, and by regarding "scientific" knowledge as what happens in that surrogate universe. Invariably, that surrogate turns out to be "too small" to accommodate organisms. In particular, this is true of the naïve and optimistic Reductionisms that have been asserted to answer the question, but which do not. In pursuing this process, I have acquired a great deal of "practical" experience about "limitations."

To me, the limitations of a method, or its inapplicability to a particular problem constitute neither a limitation of science, nor of the human mind. Nor are they inherent restrictions on the nature of the world. They mostly arise from replacing the "real world" by a small surrogate universe, a replacement that is made independently of a particular problem's exigencies. As we shall argue below, it is the essence of the *subjective* to do this.

In particular, Reductionisms in biology (and elsewhere) are adopted not because they answer the question, "What is life?", but because they emulate a method that has (sometimes) worked in dealing with inanimate nature. Reductionisms give us a small surrogate universe, consisting roughly of systems whose properties can be exhausted entirely on the basis of those of special subsystems that can be "fractionated" away from it, and studied entirely *in vitro.* These isolated fractions themselves are held to constitute a surrogate for the original system – indeed, they are the only kind of surrogate that is scientifically admissible. There are many reasons for wishing to believe that organisms fall into this class. But these reasons are irrelevant; organisms are outside such classes. I have used the word *complexity* (cf. Rosen 1991) to describe this situation. But complexity does not put organisms outside the pale of science, nor is there anything "vitalistic" about it. It is, as I have noted, simply the failure of a small surrogate universe to exhaust the real one. As such, it is essentially a mistake; an equivocation. It creates *artifacts,* as do all equivocations.

In short, a proposition like "all systems are simple (fractionable or reducible)" is not itself to be identified with "scientific knowledge." And, accordingly, failure of reductionism as a method, in the sense that there are problems to which that method does not apply, is not a limitation on science, or on scientific knowledge. Quite the contrary; we learn something basic and new about the world when it fails to be exhausted by a posited method.

II. What is "Scientific"?

To even address the question of whether or not there are "limitations of scientific knowledge," we must have an assessment of that qualifying adjective, *scientific.* It is already not so easy to characterize, or to reduce to others.

The earliest attempt to do so, or at least which I know of, goes back to Aristotle. He claimed that it was the entire job of science to account for "the why of things." This, in turn, led him to his doctrine of *causality,* in which he identified "scientific knowledge" about something (call it *X*) with all the ways of answering the question "Why *X*?" If we can say "*X* because *Y*," or that "*Y* is at least a necessary condition for the effect *X*," then this is the kind of assertion that belongs to science (though of course such an assertion may still be true or false).

Another way of saying this is that an effect *X* is entailed by its causes *Y*, i.e., by answers to question "Why *X*?" Aristotle further tacitly assumed that the totality of these answers constitute a *sufficient* condition for the effect *X*. Thus, for Aristotle, the natural world constituted a web of (at least) such causal entailments that it was the job of science to illuminate. According to this criterion, it is clear what constitutes "scientific knowledge." What it has nothing to do with is the means adopted to answer a question "Why *X*?", or with characteristics like "objectivity"; these came much later.

Partly because this Aristotelian picture of science was independent of a stipulated method for obtaining answers to such "Why?" questions, and especially because it did not focus on empirical or observational procedures (which can rarely answer "Why?" questions, in any case), that picture of science has been slowly abandoned over the past few hundred years. Indeed, the Aristotelian view, in which science is content-determined in terms of the kinds of questions it must answer, has been replaced by method-based procedures. That is, something belongs to science according to how it was obtained, not by what it is about. This constitutes a truly massive shift in outlook. Indeed, as a result of it, the question as to what is "scientific knowledge" shifts from a semantic one to one of "scientific method." In particular, "limits to scientific knowledge" has shifted from something content-based to something quite different: the adequacy of an admissible methodology.

In the sections below, we shall argue that most (if not all) questions surrounding "limits of scientific knowledge" pertain primarily to the inequivalences between the two ways we have sketched of characterizing science itself. As we have seen, the older Aristotelian view pertained to content rather than method; it consisted of answers to questions; to *information.* The second, more modern view, pertains not to content so much as to *method;* to process. The two pictures of science do not coincide. Still worse, there has never been any real consensus as to what methods are to be allowed as unquestionably producing "scientific knowledge."

III. About "Objectivity"

Probably the first thing one would say if asked to characterize "scientific knowledge" as opposed to other kinds of knowledge, is that it is *objective.*

By this is generally meant that the knowledge in question pertains to its object or referent alone, devoid of any information about how or when it was obtained.

The concept of "objectivity" already appears in the Aristotelian vision of science; it means that none of the "why?" questions about an effect *X* are answered by the processes involved in answering them. In more modern terminology, this asserts that an observation process, or an observer, plays no causal role in entailing what is observed.

However, the situation in Aristotle is confused by the presence of a causal category that he allows, but which has come to be rejected by every subsequent method-based view of what science is. That category is *finality.* Indeed, final causation has for a long time been regarded as the quintessence of subjectivity, and thus incompatible from the outset with science itself.

However, finality has no such connotation in the Aristotelian picture. In that picture, the concept of objectivity remains, even if final causation of an effect *X* is allowed; it is exactly as stated above: that *X* is simply not entailed by the process of answering "Why *X*?" That is, *X* may well have a final cause that entails it, but it is simply not the process of answering the question in that fashion. And the fact that *X* has a final cause can itself be perfectly objective, according to these criteria.

Nevertheless, the perceived need to exclude finality completely, in the name of preserving "objectivity," has been carried to ludicrous lengths in biology. For one thing, many have felt it necessary to do away with the concept of "function" entirely, one of the central concepts about organization, in general, on the grounds that it is finalistic and hence not "objective." It is astounding to watch adult physiologists twisting themselves into bizarre shapes to avoid saying things like, "the function of the heart is to pump blood." Carried still further, it is considered illegitimate for science to seek to understand anything about a part or subsystem in terms of a *larger* system or an environment with which it is interacting; hence, the Reductionistic idea that one must only seek to understand larger wholes in terms of "objective," context-independent parts, never the reverse.

In molecular biology, Jacques Monod is one of the few to actually commit to paper the frank conceptual processes at the root of it. In 1972, he states quite explicitly:

> The cornerstone of the scientific method is the postulate that nature is objective. In other words, the systematic denial that "true" knowledge

> can be got at by interpreting phenomena in terms of final causes—that is to say, of "purpose" ... science as we understand it today could not have been developed ... (without) the unbending stricture implicit in the postulate of objectivity—ironclad, pure, forever indemonstrable ... . The postulate of objectivity is consubstantial with science; it has guided the whole of its prodigious development for three centuries. There is no way to be rid of it, even tentatively or in a limited area, without departing from the domain of science itself.

In this passage, we see clearly the identification of "science" with a method, independent of any question, and the simultaneous (if subliminal) elevation of that method itself to a restriction on the material world (*"NATURE is objective"*). One need only ask if Monod's "Postulate of Objectivity" itself constitutes "scientific knowledge," and if so, about what? Indeed, to me, it is the "Postulate of Objectivity" itself that is subjective; the embodiment of a frank and clear intentionality, elevated to the status of a Natural Law. We shall have more to say about this, in a variety of contexts, as we proceed.

For the moment, let us retreat back to our original view of "objective knowledge," in the sense that acquiring it plays no role in its entailment. As we have seen, this has nothing to do with finality. But it does seem to imply a sharp dividing line between "objective knowledge" and other knowledge ("subjective knowledge"); a line based in entailment. As noted above, "subjective knowledge" cannot be entirely causally separated from the process by which it is acquired. Aristotle makes no value judgment on this division; he does not restrict or identify his view of "science" to just the "objective." This only came much later. But is there indeed such a line at all? Such a line would constitute, even for Aristotle, inherent limits to "objective knowledge." This is a deep question, which we shall discuss from a number of different angles in what follows.

IV. The Bohr-Einstein Debates

As I have argued elsewhere (e.g., Rosen 1993), physics strives, at least, to restrict itself to "objectivities." It thus presumes a rigid separation between what is objective, and thus falls directly within its precincts, and what is not. Its opinion about whatever is outside these precincts is divided. Some believe that whatever is outside is so because of removable and impermanent technical issues of formulation; i.e., whatever is outside can be "reduced" to what is already inside. Others believe the separation is absolute and irrevocable.

In either case, physics chooses a surrogate universe, bounded by criteria of "objectivity," and substitutes it for the "real" one. I will repeat here a citation (Bergman 1973) on this matter, which I have already used in *loc. cit.:*

> [Max] Planck designated in an excellent way ... the goal of physics as the complete separation of the world from the individuality of the structuring mind; i.e., the emancipation of anthropomorphic elements. That means: it is the task of physics to build a world which is foreign to consciousness, and in which consciousness is obliterated.

However interpreted, this is the "objective" world, which physics claims exclusively for itself, and which many physicists identify with "science." Anything outside can either be pulled inside ("reduction"), and hence becomes scientific as a special case of physics, or else cannot be regarded as belonging to science at all. But as we have already seen, it is already not clear whether "objectivity" is to be determined by content (the Aristotelian view) or by adherence to a method. In any case, we seem to have come a long way from science as "the why of things."

The Bohr-Einstein debates about complementarity, which occurred in the shadow of the "new" quantum theory, were precisely about what "objectivity" meant, and hence what physics itself was about. It was thus about how small (or how large) a surrogate universe it provides. A good reference about the substance of these debates are the volumes produced by Abraham Pais (1982, 1991, 1994), although Pais himself offers the opinion that, "I do not think these discussions affected the progress of physics in any way." In some trivial sense Pais is no doubt right about this (e.g., in the sense that most physicists find such issues irrelevant to what they do every day). But so much the worse for most physicists.

Both Bohr and Einstein were troubled (though in separate ways) by the wave-particle duality, revived by Einstein himself in 1905 via the photoelectric effect and his invocation of the *photon.* For over a century, it had seemed clear that light consisted of waves, on the basis of firm experimental evidence based on interference. However, beginning with the photoelectric effect, and growing ever greater as the 20th century progressed, an equally impressive body of experimental work could only be reconciled with a particulate (quantum) interpretation, and excluded waves. Indeed, some of Einstein's other work (special relativity) dispensed with the hypothetical medium *(ether)* in which waves of light had to propagate. The apparently irreconcilable wave-particle pictures, elevated to a universal status by de Broglie around 1925, paved the way for the wave mechanics of Schrödinger, and in a more roundabout way, for the matrix mechanics of Heisenberg, based on uncertainties and noncommutativities.

Bohr, who became in a sense the spokesman for new quantum theory and its interpretation through the next decades, proceeded to develop his views on complementarity to cope with the apparent contradictions between waves and particles. His position was somewhat analogous to Brouwer's position ("intuitionism") in the foundations of mathematics (an issue we will explore further subsequently). Namely, that the logical problems apparently manifested by mutually incompatible pictures really arise from trying to extrapolate classical ideas to quantum-theoretic situations. Even the idea of "mutually incompatible" is a classical idea (Brouwer said basically the same thing, in connection with the logical Law of the Excluded Middle, which held in finite realms, but not generally in infinite ones).

In particular, Bohrian complementarity argued that whether we see light as a particle or a wave was not inherent in light alone, but also depended on the way we measure or observe it. Classically, we could ignore the measurement procedure, and look at the outcome of a procedure as inherent in the light alone. But not in quantum theory. The "true" system to which the observation pertains, in a quantum realm, is the complex consisting partially of the light, but equally significant, of the apparatus or procedure that observes it.

Einstein could not accept this. He argued that something "belongs to reality" (i.e., is objective) only if it is independent of such things as how it is measured or observed. In particular, for Einstein, complementarity violates the stricture that we must not explain or describe a "real" system in terms of a particular larger one to which it might belong. This, it will be noted, is essentially Monod's "Postulate of Objectivity," that one must always look downward toward subsystems, and never upwards and outwards to a larger context.

For Bohr, then, the "objective situation" requires the specification of a means of interrogating a system, and not just the system being interrogated. The outcome of an "experiment" pertains to this larger composite system, and cannot be fractionated or separated into a part pertaining only to what is observed, and a remainder which pertains only to the means by which it is obtained. Indeed, the word "classical" pertains only to situations where such a fractionation can always be made.

So we can see that the Einstein-Bohr debates over complementary were really about what science (specifically, physics) is ultimately about; how (or even whether) objective knowledge, independent of an "organizing mind," and in which "consciousness is obliterated," can be obtained. Einstein believed that there was such knowledge, immanent alone in a thing, and independent of how that knowledge was elicited. Bohr regarded that view as "classical," incompatible with quantum views of reality, which always required specification of a context, and always containing unfractionable information about that context.

Actually, the debate over complementarity spreads far wider; it has profound causal correlates. As we saw above, Aristotelian causality involved dealing with questions or interrogatives of the form "Why X?" Bohr is in effect proposing that the variable X here is an inseparable pair, consisting of something being observed, and a procedure for describing or observing it. He is thus asserting that the original Aristotelian question, "Why X?", referring to X alone, apart from such a specific context, is ultimately meaningless, or at best can be given such a meaning only in limited "classical" realms.

In terms of surrogacy, Bohrian complementarity asserts that the classical world is too small, in itself, to provide a single coherent picture of material reality. Complementarity was his suggestion about how that world may be enlarged to accommodate quantum processes. He does so by, at root, changing the concept of "objectivity" itself from pertaining only to what is immanent entirely in a material system X to what is immanent in a system-observer pair (F, X); i.e, a larger system than X. Thus we get a bigger surrogate, which clearly cannot be "reduced" to the original "classical" world. Biological phenomena, however, raise the question whether this larger world, which is still method-based, is yet big enough. As I have extensively argued elsewhere, I do not think it is. But, as I have stated previously, this does not constitute a limitation on scientific knowledge; it is merely a failure or inapplicability of a method that has been substituted for science.

V. Some Mathematical Analogies

We have suggested above that the Aristotelian notion of causality, based on questions of the form "Why X?" establishes relations between material events, which it is the business of science to answer. In this view, the material world consists (to us) of a web of entailments between events X and Y, which take the form of such answers: "X because Y."

I have long argued that mathematics also consists of a web of entailments between propositions; *inferential* entailments. Thus, although the mathematical world is in a sense entirely subjective, in terms of its subject matter (i.e., in terms of what it is *about*), it admits the *property* of objectivity. Indeed, ironically, it is often regarded as the most objective of realms—it has often been argued that mathematical truth is the highest truth, independent alike of the external world and of the mathematician. Even God could not make $2 + 2 = 3$

Systems of inferential entailments, i.e., mathematical systems, can be compared to systems of causal entailments in many ways. For one thing, such comparisons, through vehicles of encodings and decodings between them, provide the basis of modeling relations, which I have discussed at length elsewhere (cf. Rosen 1985, 1991). But at a deeper level, mathematics, as a specimen or embodiment of entailment, can tell us things about entailment itself, and hence throw a new light on the questions we have been considering. Indeed, there are exactly parallel questions, pertaining to "limitations of mathematical knowledge," which have been discussed for a long time. These constitute what is now called the "Foundations" of mathematics, and as in science (i.e., in causal entailments) are wrapped up in arguments over whether mathematics is content-based or method-based.

Thus, mathematics has undergone periodic "foundation crises," in which discrepancies between these two views have become unbearable. We are in one today, as we shall presently discuss. For now, because of its relation to the Bohr-Einstein debates over complementarity, we shall focus on an earlier one. One of its roots was in the attempt to extend arithmetic from finite to infinite realms.

The arithmetic operations, addition and multiplication, start as binary operations (between integers, say). That is, we can only give a meaning to expressions like $(a + b)$ and $(a \times c)$; uniquely determined new integers. These operations are commutative, so we can add or multiply two integers in either order and get the same answer. Products distribute over sums, so $a \times (b + c)$ is meaningful, and is the same as $(a \times b) + (a \times c)$. On the other hand, an expression like $a + b + c$ has no immediate meaning. Or rather, it has two different ones, which can be expressed as $(a + b) + c$, and $a + (b + c)$, each of which involve only binary operations. If we set these equal, i.e., impose the associative law, then we can unambiguously remove the parentheses in these expressions, and turn addition and multiplication into a single ternary operation $a + b + c$.

It is then a theorem (cf., Chevalley 1956), proved by mathematical induction, that given any *finite* set of integers $\{a_i\}$, there is a unique sum $\sum a_i$ that is independent of the way it is parenthesized; likewise, there is a unique product, and products distribute over sums in the familiar way. In other words, these unique values are independent of how they are calculated; the values appear *immanent* in their summands or factors alone; they are *objective.*

In the sixteenth century, mathematicians exuberantly began to explore what happens if this restriction of finiteness is removed, i.e., if we consider sums with an infinite number of summands, or products with an infinite number of factors. At first, in the hands of people like Euler and the Bernoullis, it seemed as if a whole new world of ineffable beauty was emerging, populated by relations like

$$1 + \frac{1}{9} + \frac{1}{25} + \cdots + \frac{1}{(2n-1)^2} + \cdots = \frac{\pi^2}{8},$$

or

$$\frac{2}{1} \times \left(\frac{4}{3}\right)^{\frac{1}{2}} \times \left(\frac{6}{5} \times \frac{8}{7}\right)^{\frac{1}{4}} \times \left(\frac{10}{9} \times \frac{12}{11} \times \frac{14}{13} \times \frac{16}{15}\right)^{\frac{1}{8}} \times \cdots = e.$$

At first, these results appeared just as "objective" as finite arithmetic. But then apparent absurdities began to creep in, slowly at first, then in a torrent. The worst of them was perhaps

$$1 - 1 + 1 - 1 + \cdots = ?,$$

which seemed to entail the ultimate absurdity, $1 = 0$.

So something was clearly wrong here, a Foundation Crisis. There seemed to be only three choices: (a) to restrict arithmetic to finite realms exclusively, where we were safe, (b) to venture into infinite realms, but to find a way to choose our factors and summands judiciously enough to retain "objectivity," or (c) to somehow enlarge the idea of "objectivity" itself.

As everyone knows, that Crisis was (temporarily) resolved by Cauchy around 1805 on the basis of (b). He introduced the notion of *convergence,* and a number of algorithmic criteria for deciding whether a given sequence or series converged. Convergence, in this sense, was a generalization of the "objectivity" automatically enjoyed in finite realms; the *value* of the limit of a convergent sequence was independent of how it was obtained, and inhered in the sequence alone. This stopgap rescued mathematical analysis as a secure discipline, and even had profound ramifications in theoretical physics, through the notion of the derivative (velocity) as a well-defined variable, independent of how it was evaluated or measured. Without this, there could be no notion of *phase.*

A series like the last one displayed above is called *divergent,* and was regarded as meaningless; beyond the realm of mathematics itself. Such a sequence was thought of as having *no limit.* On the other hand, in a precise sense, "most" sequences thus diverge; only very special ones satisfy the Cauchy criteria. In fact, what happens with divergent sequences is that they have *many* distinct limiting values; values that depend precisely on how they are evaluated, and *not just* on the sequence alone.

Thus, a "limiting value" of a divergent sequence σ is a pair, which we may denote by (F, σ), where F is a specific evaluation process. In a sense, this is a perfectly objective number, determined by that pair. For instance, in the example of divergence given above, in which

$$\sigma = 1 - 1 + 1 - 1 + \dots ,$$

the apparent absurdity $1 = 0$ arises from not specifying F; or rather, by tacitly using two different Fs, and then equating the results, which we cannot do.

I believe the reader can immediately see the parallels between this situation and that involved in the Bohr-Einstein debates. In effect, Einstein was arguing that only convergent situations, in which limits are independent of how they are evaluated, are entitled to be called "real" or objective. It was Cauchy's position that only convergent sequences are allowable. Bohr's position was, rather, that the material world, in its quantum aspects, was like divergence; perfectly objective, but requiring further "information," pertaining to how it is evaluated, to determine a specific one of many complementary limits the sequence may have.

Let us pause to recapitulate what we have just said, since it is important. We were considering familiar operations of arithmetic, like addition, on ordinary numbers. We observed:

a. If we restrict ourselves to finite numbers of summands, a sum is uniquely determined, and independent of the way in which it is computed or evaluated. If we identify this independence with "objectivity," then this objectivity is in fact entailed by finiteness. That is, if σ denotes a finite set of summands, then it is a theorem that $F(\sigma) = G(\sigma)$, where F and G are distinct evaluation processes.
b. If we go to infinite realms, there is nothing to entail this "objectivity." It is no longer generically true that $F(\sigma) = G(\sigma)$ when σ is infinite.

c. If we generalize the "objectivity" condition $F(\sigma) = G(\sigma)$, we obtain a world of convergent sequences. This is, however, no longer a theorem, as it was before; rather, it is something *imposed,* or *presupposed,* in order to demarcate a world of infinite sequences that still manifest a property familiar from finite sums.
d. We then go on to identify the world so demarcated with "mathematics." What is left out of that world (which is, in fact, "most" sequences) is regarded as "nonobjective," since a limiting value depends on the choice of an evaluation procedure F, and not on a sequence σ alone.
e. Thus, what we call "mathematics" in this procedure becomes inherently method-dependent, and not content-dependent. As such, it has (pardon the pun) inherent limits.
f. Nevertheless, the only real subjectivity here is entirely in the decision to replace "arbitrary sequence" by "convergent sequence," and to replace a pair (F, σ) by σ alone. This choice is not mandated or entailed by anything in the content of mathematics itself. Rather, it is motivated by an intention, to retain customary habits, familiar from the finite realm, as intact as possible in the infinite one.

VI. A Word About Computability

Another good example of the phenomenon we have just described comes from our current mathematical Foundation Crisis, rather than from a previous one. I have written much about it in the past, so I will cover it quickly. Although its immediate historical roots are somewhat different, it is in fact rather closely related conceptually to the issue of convergence. It has to do with *computability,* viewed again as a generalization of finiteness.

It was long a hope, or expectation, that the entailment processes of mathematics could be equivalently replaced with word-processing algorithms. Mainly, the hope was that in such severely restricted, finitely-generated contexts, consistency of arbitrary mathematical theories, like set theory, could be *guaranteed; entailed.* This was the world of Formalism. It turned out, of course, that "most" mathematical systems were not formalizable; this was the upshot of the venerable Gödel Theorem, which has (among many other things) precipitated these workshops on "limits to scientific knowledge."

The theorem actually compares a method-based simulacrum of mathematics, based on formalizability (computability) with a content-based concept, and shows that the former is inevitably much smaller than the latter. The problem here is in a way much more acute than simply replacing "sequence" by "convergent sequence," as Cauchy did; it, in effect, sharply limits the evaluation processes (what we called F before) which allows limits of sequences to be evaluated at all. So the formalist world is very impoverished from the outset. In fact, it is limited to what can be done entirely by the iteration of rote processes; a limitation which will not carry you from the finite to the infinite, nor back again.

In this formalistic world, "objectivity" is interpreted yet again, as entailment arising from purely syntactic rules. Roughly, something is "objective" if and only if it could be carried out by a properly programmed machine; i.e., as a matter of software and hardware. This conception spills over into science, and especially into biology, by recasting Monod's dictum ("NATURE is objective") into the still more restrictive form "NATURE is computable."

Indeed, it might be noticed that Bohrian complementarity would be a hard thing to convey to a finite-state machine. Part of the problem here is that such a machine is, in material terms, entirely a classical device, both in its hardware and in its software; all it could see of a non-classical, quantum realm is noise.

VII. On Complexity

As I remarked at the outset, my interest in the problems dealt with above is a consequence of my scientific concerns with biology, and particularly with the question, "What is life?", the central question of biology. It became clear to me that Reductionism was not the way to find an answer. Indeed, the Laplacian *Geist,* the Reductionistic ideal, would make an extremely poor biologist. So too would his quantum-mechanical analog.

Since I did not regard myself as method-bound, I explored alternatives, especially under the rubric of "relational biology" (cf. Rosen 1991). I ended up with a class of systems ((M, R)-systems), which seemed to me to be (a) perfectly "objective," but (b) fell outside the category of "mechanisms," and could not be understood in terms of Reductionistic method alone. I cannot claim that these (M, R)-systems fully answer the question "What is life?", but I do claim that the answer must at least comprehend them.

The (M, R)-systems manifest inherent semantic properties, expressed in closed causal loops within them. Such closed loops, in inferential contexts, are called *impredicativities.* It is such impredicativities that Formalism rejects as "subjective," and which scientists like Monod *(vide supra)* claim are inconsistent with science itself, at least as they comprehend it

At any rate, my little (M, R)-systems are inherently unformalizable as mathematical systems. That means: not only do they have noncomputable models, but any model of them that is computable is not itself an (M, R)-system, and hence misses all of its biology.

You cannot "reduce" nonformalizability or noncomputability to formalizability. That is simply a fact; it is not a limitation. Likewise, you cannot, for example, "reduce" nonexact differential forms to single (potential) functions; again, that is a fact, and not a limitation. Nor can you "reduce" divergent sequences to convergent ones. Indeed, such "reducibility" is in itself a rare and nongeneric phenomenon; that too is a fact; a fact as "objective" as anything.

I originally called a (material) system *complex* if it had noncomputable models; otherwise, *simple.* People like Monod simply postulate that every *material system is simple;* i.e., there are no complex systems in nature, and identify this postulation with science itself. As noted above, this makes science method-based, rather than content-based, and inevitably limits its content only to what is consistent with the postulated method. This has always turned out to be a Procrustean bed, which inevitably creates limits to the science (or to the mathematics) so circumscribed.

This is how I view the Gödel theorem. It exhibits "mathematics" (or at least number theory) as a profound generalization of formalizability (or, alternatively, reveals formalizability to be a most severe specialization of mathematics). Mathematical rigor (i.e., "objectivity") does not reside in finitely-based syntactic rules alone. In my previous terminology, Gödel's Theorem demonstrated the *complexity* of number theory—its irreducibility to simple formalizations. I believe that biology does the same thing for our contemporary views of physics.

Indeed, one of the upshots of Monod's "Principle of Objectivity" is that one must never claim to learn anything new about physics (i.e., about matter) from a study of organisms; or, stated another way, that an observer would see exactly the same universe, governed by the same laws, whether life existed in it or not; the difference between them is a conceptually trivial difference in *state.* A mathematical analog of this assertion is that, e.g., one must never claim to learn anything new about set theory from a study of, e.g., groups. I believe that, in mathematics at any rate, Gödel's Theorem already refutes any such claim.

Set theory is currently the mathematical version of the physicist's search for a "theory of everything"; a theory from which everything is inferentially entailed. As noted above, in earlier centuries, the physicists' dream was embodied in the Laplacian *Geist,* who could know the motion of every particle and every force; who could formulate and solve every N-body problem. He might not be able to tell the difference between a universe with life in it and one without it; but then he could not conceive of a system which was not an N-body system. The limitations to what he could know are limitations in *him,* not in the universe he perceives; and he could never even know what they are.

I suggest that we humans are more fortunate than the hypothetical Ultimate Reductionist, in our ability to perceive complexity. That is, to recognize the necessity to pull ourselves outside the limitations of self-imposed methodologies, which create nonexistent "limits" to knowledge itself.

I believe, in short, that Aristotle was more correct in his view of science as a content-based thing, than the more currently orthodox views of science as constricted by a method. If this is so, then the problem of "limits" to science evaporates into mist.

Appendix: On Emergence

As a biologist, I have been engaged in various ways in a long-standing debate concerning the scientific status of the concept of "emergent novelty." The debate itself is of very broad currency, but it is perhaps sharpest in biological contexts. It touches directly on the issues we have been discussing; moreover, it is illuminated by the remarks made above. Hence it appears worth discussing the issue separately, if briefly, in this Appendix.

In a rough intuitive sense, "emergence" pertains to situations in which a system is, in some sense, "more than the sum of its parts." Another way of saying this is that a study of "parts" does not, by itself, entail properties of the whole from which the parts have been fractionated away. In the language we have used above, the "parts" do not suffice to answer all the "Why?" questions we can ask about the whole; those parts are not sufficient to entail the whole.

One can immediately see why methods of Reductionism, based precisely on the isolation and study of such "parts," look upon concepts of emergent novelty with great hostility. They tend to denounce the concept itself as unscientific and metaphysical.

On the other hand, the whole idea of "complexity" has always been tied up with the concept of emergence. For instance, von Neumann early argued that there was a "threshold of complexity" (I would prefer to use the word "complication"), below which automata (e.g., neural networks) could only deteriorate, but above which entirely new capabilities (learning, development, growth, reproduction, evolution) emerge. These are, of course, *biological* capabilities; pushing a material system across this threshold of complexity amounts to creating life. Indeed, in this view, the emergent property of "complexity" itself becomes a causal entity, a way of answering "Why?" questions: "Why is this system alive?" Because it is "complex"; over that presumptive threshold.

In the light of these considerations, let us reconsider the example of divergent sequences introduced in Section V above, and analogized with Bohrian complementarity. Clearly, from the standpoint of finite arithmetic, divergence is itself an emergent property. The dependence of a limit of an infinite sum on the way its summands are ordered and parenthesized is something without a counterpart in finite sums, and is unpredictable from them *alone.* Or, stated another way, finite arithmetic does not provide enough entailment to answer "Why?" questions about limits of divergent sequences.

This kind of situation is characteristic of emergence itself; a paucity of entailment in a world of "parts," and hence an inability to answer "Why?"-questions about systems from which the "parts" have been isolated.

As we have already seen above, if you make no concession to the additional modes of entailment on which the emergence itself depends, you get $1 = 0$.

References

[1] Chevalley, C. 1956. *Fundamental Concepts of Algebra.* Columbia University Press, New York.

[2] Monod, J. 1971. *Chance and Necessity.* Knopf, New York.

[3] Pais, A. 1982. *Subtle is the Lord.* Clarendon Press, Oxford.

[4] Pais, A. 1991. *Niels Bohr's Times.* Clarendon Press, Oxford.

[5] Pais, A. 1994. *Einstein Lived Here.* Clarendon Press, Oxford.

[6] Rosen, R. 1991. *Life Itself.* Columbia University Press, New York.

[7] Rosen, R. 1993. *Theoretical Medicine,* 14, 89–100.

Chapter 9

UNDECIDABILITY EVERYWHERE?

Karl Svozil

I. Physics After the Incompleteness Theorems

There is incompleteness in mathematics [1–7]. That means that there does not exist any reasonable (consistent) finite formal system from which all mathematical truth is derivable; and there exists a "huge" number [8] of mathematical assertions that are independent of any particular formal system. That is, these statements – as well as their negations – are compatible with the standard formal systems of mathematics. Take, for example, the Continuum Hypothesis or the Axiom of Choice. Both are independent of the axioms of Zermelo-Fraenkel set theory.

Can such formal incompleteness be translated into physics or the natural sciences, in general? Is there some question about the nature of things that can be proved to be unknowable for rational thought? Is it conceivable that the natural phenomena, even if they occur deterministically, do not allow their complete description?

Of course it is! Suppose there exists free will. Suppose further that an observer could predict the future. Then this observer could freely decide to act in such a way as to invalidate that prediction. Hence, in order to avoid paradoxes, one has either to abandon free will or accept that perfect and complete prediction is impossible.

The above argument may appear suspiciously informal. Yet, it makes use of the diagonalization technique, which is one of the royal roads to a constructive, rational understanding of undecidability in the formal sciences. What Gödel and others did was to encode the argument in a language suitable to their area of research. To translate and bring similar issues into mainstream natural science is, at least in the author's opinion, the agenda of present concern on rational limits to scientific knowledge.

Before discussing these questions further, we should first clarify our terms. By a *physical phenomenon,* we shall understand an event that is (irreversibly) observed. A typical example of a physical phenomenon consists of a click in a particle detector; there can be a click or there can be no click. This Yes/No scheme constitutes experimental physics in a nutshell (at least according to a theoretician). All empirical evidence that we will ever have is acquired from this type of elementary observation.

Physical theories purport to relate to the physical phenomena. Taken at face value, they consist of phenomena themselves. As observers, we would not be able to know about these theories if we did not observe their representation or code. A typical code of, say, the theory of electrodynamics, consists of letters and symbols printed in a book on electrodynamics. Reading these symbols constitutes an act of observation.

Why should anyone bother about the intrinsic representation of physical entities such as observations and theories? The answer is that this issue is crucial for understanding undecidability. Gödel, for instance, proved his incompleteness theorems by successfully coding a (generic) theory about arithmetic *within* arithmetic.

At some point, we have to confront the question, "Is there a physical correspondent to the notion of logical inconsistency; that is, to a logical contradiction?" [9]. Can a particle, for example, be here and somewhere else (i.e., not here) at the same moment [10]? On the phenomenological level, the answer is No. To put it pointedly, there is no such thing as an inconsistent physical phenomenon. In a Yes/No experiment with two possible outcomes, only one of these outcomes will actually be measured. In contradistinction, a *theoretical description* might allow the consistent "existence" of mutually exclusive states if it is indeterministic (probabilistic). We shall come back to these issues later.

The term *indeterminism* stands for any process that cannot be described finitely and causally, that is, by an algorithm. As a metaphor, we may say that in indeterministic physical systems, "God plays dice." By definition, indeterminism implies undecidability. If there is no cause, there cannot be any predictable effect. That is the whole story. Period.

Let us be more specific and consider two examples. Firstly, in quantum mechanics, the conventional probabilistic interpretation of the wave function contends that it is a complete description. Single outcomes cannot always be deterministically accounted for. This is the quantum dice. Secondly, in the case of "deterministic chaos," the (Martin-Löf/Solovay/Chaitin) randomness of "almost all" initial conditions represented by elements of the classical mechanical continuum is successively recovered during the time evolution of the system—bit by bit. Therefore, if one believes in the quantum dice and in the physical reality of the classical continuum, then, by definition, there must be undecidability in physics.

Why can't we simply sit back, relax and stop here? We have just demonstrated the fact that present-day physical theories contain indeterministic features that evade any kind of complete prediction. Why isn't this the end of the story? The trouble is that we shall never be sure that the probabilistic interpretation of the wave function is complete. Nor do we know whether or not the classical continuum is physically meaningful [11]. There might be a "secret arena" hidden from us, in which everything can be deterministically accounted for. If we give up now and uncritically accept indeterminism as a matter of unquestionable fact, then we may be heading for trouble. Indeterminism, as it is promoted by the physics community at the *fin de siécle (millénaire),* might be a degenerative research program.

So it does not seem inappropriate to try to reinterpret physical indeterminism constructively. In doing so, it is necessary to study in more detail undecidability and incompleteness for systems that are mechanistic. By *mechanistic* we mean that they are finitely describable and causal in all of their aspects. (In what follows, the terms *mechanistic, algorithmically, computable* and *recursive* are used synonymously.) In mechanistic systems, every effect has a cause. But—if everything has a cause, if everything can be deterministically accounted for, computed and predicted—how can there be incompleteness?

The answer derives from a most important epistemological issue: Although in principle every effect may have a cause, such causes might not be knowable by intrinsic observers, i.e., observers that are *inside* the system. Here we are introducing an inside-outside distinction. *Intrinsic observers* are embedded in the system they observe. The system is their "Cartesian prison" [12–19]; they cannot step outside it. Intrinsic observers are bound to observations that are intrinsically operational [20]. They can only refer to intrinsic entities [21] (cf. the "virtual backflow" [22]); their theories must be intrinsically codable.

Rather then attempting a formalization of intrinsic perception (cf. [19]), let us consider a metaphor, or rather a nightmarish virtual reality scenario, a Zen *koan:* Suppose that we are thrown into a prison (or, perhaps, a lunatic asylum) without any explanation. In this prison we see people animatedly talking to each other. Yet we neither understand their language nor the reasons, rules and laws of that establishment. In a metaphorical sense, the real world can be perceived as just such a prison. And science might be simply one attempt among many to make sense out of the situation.

But let's continue with determinism. The way it was defined, a mechanistic physical system corresponds in a one-to-one fashion to a process of computation. This computation can, for instance, be implemented on a universal cellular automaton, a universal Turing machine or any other universal computer. This computer, in turn, has a one-to-one correspondence with a formal system of logic. With these two identifications, namely *mechanistic physical system* ≡ *computation* ≡ *formal system,* we bridge the gap to formal undecidability, and the term *system* could stand for any of these three entities [19, 23].

We have now set the stage. Let us summarize where we stand. We would like to consider mechanistic physical systems. Intrinsic observers are embedded within such systems. These intrinsic observers register physical phenomena that are operational. Moreover, they develop theories that are intrinsically codable. Our aim is to determine whether or not certain physical phenomena and theoretical propositions become undecidable, under these constraints. In doing so, we must translate the notion of recursion-theoretic undecidability into physics. Our translation guide will be the equivalence between mechanistic physical systems, computation, and formal systems.

II. Physics after Turing

We should be quite clearly aware of the fact that there is no possible formalization of undecidability other than recursive function theory and formal logic. If one objects to the idea of logical or computer science terminology creeping into physics, then there is no room for discussion of these issues. Undecidability in physics marks the integration of yet another abstract science—recursion theory—into physics.

Gödel himself did not believe in the physical relevance of his incompleteness theorems. In particular, he rejected the notion that they had anything to say about quantum mechanics [24]. One might speculate that Gödel had been brainwashed by Einstein, who was adamantly opposed to the Copenhagen interpretation of quantum mechanics. Einstein thought that the Copenhagen interpretation, with its negation of the existence of the results of unperformed measurements, was a degenerative research program [25–26], a fact well expressed by his legendary *dictum,* "God does not play dice."

Yet there *is* a straightforward extension of formal incompleteness to physics. It is based on Turing's proof that certain propositions about universal computers, which basically mimic "mechanical" paper-and-pencil operations on a sheet of paper, are undecidable. In particular, it is impossible to predict whether or not a particular computational task on a universal computer will eventually terminate and output a particular result. Therefore, if we construct a physical device capable of universal computation, there are some propositions about the future of this system that are provably undecidable.

Let us be more specific and prove (algorithmically) this assertion about undecidability. This result is often referred to as the "halting theorem" or the "recursive undecidability of the halting problem." We will use the technique of *diagonalization,* which was pioneered by Cantor in a proof of the nondenumerability of the real numbers. This technique is the most useful single tool at our disposal for exploring the undecidable.

The strategy of diagonalization is to assume a statement whose existence should be disproved. Then by manipulation of the statement, we derive a paradox, a contradiction. The only consistent way to avoid this paradox is then to abandon the original statement. For the purpose of a formal proof, any paradox can, in principle, be used, as long as it is codable into formal entities. Gödel [1], as well as Turing [3], used "the liar paradox" [27] for their incompleteness theorems. Gödel was well aware of the fact that almost any classical paradox might do equally well.

We shall prove the recursive unsolvability of the halting problem algorithmically. That is, we shall use informal terminology, which by the Church-Turing thesis [28–29] is assumed to correspond to certain formal expressions.

Assume, for the moment, that there is a mechanistic way to foresee whether or not a particular computation will terminate; or, equivalently, if the computation will output a particular string of symbols and then terminate. Consider an arbitrary program $B(x)$ whose input is a string of symbols x. Assume there exists a "predictor," called `PREDICT`, that is able to decide whether B terminates when processing x or not. Using the predictor `PREDICT`$(B(x))$, we shall construct another computing agent A, whose input is any effective program B and that proceeds as follows: Upon reading the program B as input, A makes a copy of B. This can be easily done, since the program B is presented to A in some encoded form, i.e., as a string of symbols. Next, the program A uses the code of B as the input string for B itself, i.e., A forms $B(B)$. The program A now hands $B(B)$ over to the prediction subroutine `PREDICT`.

Program A now proceeds as follows: If `PREDICT`$(B(B))$ states that $B(B)$ halts, then A does not halt. This can be realized, for instance, by an infinite `DO`-loop. If `PREDICT`($B(B)$) claims that $B(B)$ does *not* halt, then A does halt. (This is the diagonalization step.) We shall now confront the program A with a paradoxical task by choosing A's own code as input to itself. Notice that B is arbitrary and has not yet been specified. The deterministic program A can be represented by an algorithm with the code of A. Therefore, it is possible to substitute A for B. Assume that A is restricted to classical bits of information. Then, whenever $A(A)$ halts, `PREDICT`($A(A)$) forces $A(A)$ not to halt. Conversely, whenever $A(A)$ does not halt, the predictor `PREDICT`($A(A)$) steers $A(A)$ into the halting mode. In either case, one arrives at a contradiction. Classically, this contradiction can only be consistently avoided by assuming the nonexistence of such a program A and, since the only nontrivial feature of A is the use of the predictor subroutine `PREDICT`, the impossibility of the existence of any such universal predictor.

Notice that the above argument is nothing but a rephrasing of the informal argument against free will or complete predictability given at the beginning of this chapter! Popper [30] considered these issues already in the 1940s. More sophisticated models have been put forward by Wolfram [31], D. Moore [32], and da Costa and Doria [33]. These approaches essentially embed a universal computer (or an equivalent system of Diophantine equations) into a classical physical structure like a field. The system is assumed to be infinite to allow for an infinite tape or its equivalent. Undecidability then follows from the recursive unsolvability of the halting problem.

In short, reasonable (consistent) theories predicting the future behavior of arbitrary mechanistic physical systems are impossible. So if one believes in the physical relevance of the model of universal computers, then no physical theory can predict all physical phenomena. In particular, there are certain physical prediction tasks that are undecidable.

But what if one insists that any computation should remain finite? Then, in principle, it would be possible to construct a predictor that would have to simulate the system long enough and fast enough to complete the prediction. Could such a prediction be done in a short enough time to be useful? And what if a finite predictor tries to predict itself? These questions take us back to more involved quantitative issues, which we shall now attack.

III. The Busy Beaver Function: Watching a System Explode

The Busy Beaver function [34–35] addresses the following question: Given a Turing machine program whose description is finite, more precisely, not greater than n bits in length, what is the largest number $\Sigma(n)$ that such a program can produce before halting (or, alternatively, before recurring to the system's original state)?

A related question is: What is the upper bound on the running time (or, alternatively, recurrence time) of a program of length n bits before it terminates? An answer to this second question gives us an indication of how long we have to wait for the most time-consuming program of length n bits to terminate. That, of course, is a worst-case scenario. Many n-bit programs will halt before the maximal halting time.

Let us denote the maximal halting time by `TMAX`. Knowledge of `TMAX` "solves" the halting problem quantitatively, since if we knew that maximal halting time, then for an arbitrary program of n bits we would have to wait just a little bit longer than `TMAX`(n). If it were still running, then we could be sure that it would run on forever; otherwise, it would have already halted. In this sense, knowledge of `TMAX` is equivalent to possessing a perfect predictor `PREDICT`. But since we have seen that the latter does not exist, we may expect that `TMAX` cannot be a constructive function.

Indeed, Chaitin proved [35–36] that `TMAX`$(n) = \Sigma(n + O(1))$ is the shortest time by which all halting programs of size less than or equal to n bits have actually halted. For large values of n, $\Sigma(n)$ grows faster than any computable function. More precisely, let f be an arbitrary computable function. Then there exists a positive integer k such that $\Sigma(n) > f(n)$ for all $n > k$.

It's easy to see that any system trying to evaluate the Busy Beaver function "blows up." Originally, Rado [34] asked how many 1s a Turing machine having n possible states and an empty input tape could print on that tape before halting. The first values of the Turing Busy Beaver function, $\Sigma_T(n)$ are finite and, in fact, known [37–38]:

$$\Sigma_T(1) = 1, \qquad \Sigma_T(2) = 4,$$
$$\Sigma_T(3) = 6, \qquad \Sigma_T(4) = 13,$$
$$\Sigma_T(5) \geq 1,915, \qquad \Sigma_T(7) \geq 22,961,$$
$$\Sigma_T(8) \geq 3 \cdot (7^{92} - 1)/2.$$

What does all this mean for physics? One consequence is that for mechanistic (but unbounded) systems describable by n bits, the recurrence time grows faster than any computable function of n. This shows that we would have to wait indeed a "very long" time in order to make sure that any algorithm of given length that eventually halts would have actually done so!

IV. Is Complete Self-Description Possible?

Any causal prediction requires a theory of the system that one wants to predict. In the intrinsic-observer scenario described above, there is no way to separate the observer from the system. We have to deal with self-description.

Can observers embedded in a system ever hope for a complete theory, or self-description? Let us rephrase the question. Is it possible for a system to contain a "blueprint," that is, a complete representation, of itself? This issue was raised by von Neumann in his investigations of self-reproducing automata. Indeed, he showed that an automaton can reconstruct a perfect replica of itself—provided such a "blueprint" exists [28–29, 39].

To avoid confusion, it should be noted that it is never possible to have a finite copy of a system within the system itself as proper part. The trick is to employ *representations* or *names* of objects, whose code can be smaller than the objects themselves and can indeed be contained in that object (cf. [29, p. 165]). Gödel's first incompleteness theorem is just such an example. Any book on electromagnetism or cellular DNA are others.

A completely different issue is how such a theoretical self-description is obtained. Here we have to make a distinction. As in the above case, a complete theory, or self-description, might be obtained *passively* from "intuition," "God" or an "oracle." (Of course, one could never be sure that it is the right one.) But it is generally impossible for an intrinsic observer to *actively* examine its own system and to thereby construct a complete theory. One reason for this is self-interference and complementarity, as will be described further below. Another reason is the recursive unsolvability of the rule-inference problem, as discussed next.

As a comfort to those who see self-reference in the "Cartesian prison" as the source of all problems, consider the very nice theorem by Gold [40]. This result is sometimes referred to as the recursive undecidability of the rule inference problem: For any mechanistic intelligence A, there exists a total recursive function f such that A does not infer f. In more physical terms, there is no systematic way of finding a deterministic law from the input/output behavior of a mechanistic physical system.

A way to (algorithmically) prove Gold's Theorem uses the Halting Theorem. Suppose that it is indeed possible to derive laws systematically. Let us call this agent or computable function `RULE`. The function `RULE` would have to watch the behavior of the system it analyzes. But any complete analysis would require the observation until time `TMAX`(n), where n is the minimal description size of the system. Since `TMAX`(n) grows faster than any computable function of n, the function `RULE` cannot be computable.

So, even if we were in a "God-like" position and could be disentangled and freed from the observed system, we would have to cope with the fact that there is no systematic way of deriving causal laws. Indeed, we may safely state that, except for the most elementary phenomena, deriving causal laws remains a rare art!

V. Predicting Predicting

Of what use is a complete theory? Is it possible for an observer in a finite amount of time to predict its own state completely?

An intuitive understanding of the impossibility of complete self-comprehension can be obtained by considering a variant of Zeno's paradox of Achilles and the Tortoise (called the "paradox of Tristram Shandy" by Popper [30]): In order to predict oneself completely, one has to predict oneself predicting oneself completely, and then one has to further predict oneself predicting oneself predicting oneself completely, then one has to ... . The infinite regress is clear.

VI. Intrinsic Undecidability

The Hindu notion of Maya suggests that the world of the senses is illusory, that observations are distractive. Plato's cage metaphor emphasizes the distinction between objects and what we may be able to observe about these objects. One day, complementarity might be perceived as a variation of this ancient theme.

Hardly any feature of quantum mechanics has given rise to as many speculations as *complementarity.* Intuitively, complementarity states that it is impossible to (irreversibly) measure certain observables simultaneously with arbitrary accuracy. The more precise one of these observables is measured, the less precise can be the measurement of other—complementary—observable. Typical examples of complementary observables are the direction of spin of one particle measured at different angles (mod π) [41–42].

The intuition (if intuition makes any sense in the quantum domain) behind this aspect of measurement is that the act of (irreversible) observation of a physical system causes a loss of information by (irreversibly) interfering with the system. Thereby, the possibility to measure other aspects of the system is destroyed.

But this is not the whole story. Indeed, there is reason to believe that—at least up to a certain magnitude of complexity—any measurement can be "undone" by a proper reconstruction of the wave function. A necessary condition for this to take place is reversibility of the measurement, such that after the state reconstruction, *all* information about the original measurement is lost. Schrödinger, the creator of wave mechanics, liked to regard the wave function as a sort of *prediction catalog* [43]. This catalogue contains all potential information. Yet, it can be opened only at a *single* page. The prediction catalog may be closed before this page is read, in which case it could then be opened once more at another, complementary, page. But there is no way an observer can open the catalogue at one page, read it and then (irreversibly) memorize the page, close the catalogue and then open it at another, complementary, page. (But note that two non-complementary pages that correspond to two co-measurable observables can be read simultaneously.)

This may sound a little bit like voodoo. It is tempting to speculate that complementarity can never be modeled by classical metaphors. Yet, classical examples abound. A trivial one is a dark room with a ball moving in it. Suppose we want to measure its position and its velocity. We first try to measure the ball's position by touching it. This finite contact inevitably causes a finite change of the ball's motion. Therefore, we can no longer measure the initial velocity of the ball with arbitrary position.

There are a number of more faithful classical metaphors for quantum complementarity. For instance, consider Cohen's "firefly-in-a-box" model [44], Wright's urn model [45], or Aerts' vessel model [46]. In what follows, we are going to explore a model of complementarity pioneered by E. F. Moore [47]. It is based on extremely simple systems—probably the simplest systems you can think of—a finite automaton. The finite automata we consider here are objects having a finite number of internal states and a finite number of input and output symbols. Their time evolution is mechanistic, and can be written down as tables in matrix form. There are no built-in infinities anywhere; no infinite tape or memory or nonrecursive bounds on the run time.

Let us develop *computational complementarity,* as it is very often called [48], as a game between you as the reader and me as the author. The rules of the game are as follows. I first give you all you need to know about the intrinsic workings of the automaton. For example, I tell you that "if the automaton is in state 1 and you input the symbol 2, then the automaton will make a transition into state 2 and output the symbol 0," and so on. Then I present you with a black box containing a realization of the automaton. The black box has a keyboard with which you type-in the input symbols. It has an output display on which the output symbols appear. No other interfaces are allowed. Suppose that I can choose the initial state of the automaton at the beginning of the game, but I do not tell you this state. Your goal is to discover this state by experimenting with the automaton. You can simply guess or rely on your luck by throwing dice. But you can also perform clever input/output experiments and analyze the experimental data in order to uncover the initial state. You win the game if you give the correct answer; I win if you guess incorrectly. (So I have to be mean and select worst-case examples.)

Suppose you try very hard. Is being clever sufficient to win the game? Will you always be able to uniquely determine the initial automaton state?

The answer to that question is "No." The reason is that there may be situations when the input causes an irreversible transition into a state that does not allow any further queries about the initial state. This is the correspondent of the term "self-interference" mentioned above. Any such irreversible loss of information about the initial state of the automaton can be traced back to many-to-one operations [49]: different states are mapped onto a single state with the same output. Many-to-one operations, like the deletion of information, are the only source of entropy increase in mechanistic systems [49–50].

In the automaton game discussed above, one could, of course, restore reversibility and recover the automaton's initial state by Landauer's "Hansel and Gretel"-strategy. This strategy consists of introducing an additional marker at every many-to-one node indicating the state before the transition. But then, as the combined automaton/marker system is reversible, going back to the initial state erases all previous knowledge. This is analogous to the re-opening of pages of Schrödinger's prediction catalog.

This might be a good moment to introduce a simple example. Consider an automaton that can be in one of three states, denoted 1, 2 and 3. This automaton accepts three input symbols, namely 1, 2 and 3, but outputs only two different symbols, namely 0 and 1. The transition function of the automaton is as follows: given input 1, it makes a transition to (or remains in) state 1; presented with input 2, it makes a transition to (or remains in) state 2; given input 3, it makes a transition to (or remains in) state 3. This is a typical irreversible many-to-one operation, since a particular input steers the automaton into that state, regardless of which of the three possible state it was in previously. The output function is also many-to-one and rather simple: whenever both the current state and the input coincide—that is, whenever the guess was correct—it outputs 1; otherwise, it outputs 0. So, for example, if the automaton was in state 2 or state 3 and receives input 1, it outputs 0 and makes a transition to state 1. There it awaits another input. These automaton specifications can be conveniently represented by diagrams such as the one drawn in Figure 1.

Complementarity manifests itself in the following way. Assume again that you do not know the initial state of the automaton. The goal is to find it. Let us assume that you input 1. Then the automaton responds with 0 or 1, depending on whether it has been in the initial states 2 or 3 (corresponding to output 0), or 1 (corresponding to output 1). Therefore, the experiment induces a partitioning of the set of automaton states. In the above case, the partitions are $\{1\}$ and $\{2, 3\}$. The set of partitions can be denoted by

$$v(1) = \{\{1\}, \{2, 3\}\}.$$

Likewise, input of the symbols 2 and 3 induces the following partitions

$$v(2) = \{\{2\}, \{1, 3\}\}, v(3) = \{\{3\}, \{1, 2\}\},$$

respectively. After this irreversible measurement, the automaton is surely in the state corresponding to the input. Hence, if you input 1, you measure the observable corresponding to the proposition "the automaton is in state 1" (or, equivalently, "the automaton is not in state 2 or 3"). But you lose the option to experimentally measure the observables corresponding to the proposition "the automaton is in state 2" and "the automaton is in state 3." Likewise, if you input 2 (3), you measure the observable corresponding to the proposition "the automaton is in state 2 (3)" (or, equivalently, "the automaton is not in state 1 or 3 (2)"). But you lose the option to experimentally measure the observables corresponding to the proposition "the automaton is in state 1" and "the automaton is in state 3 (2)." At the beginning of each experiment

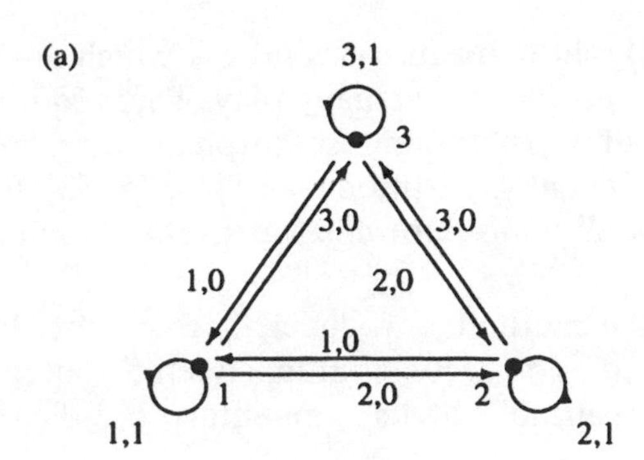

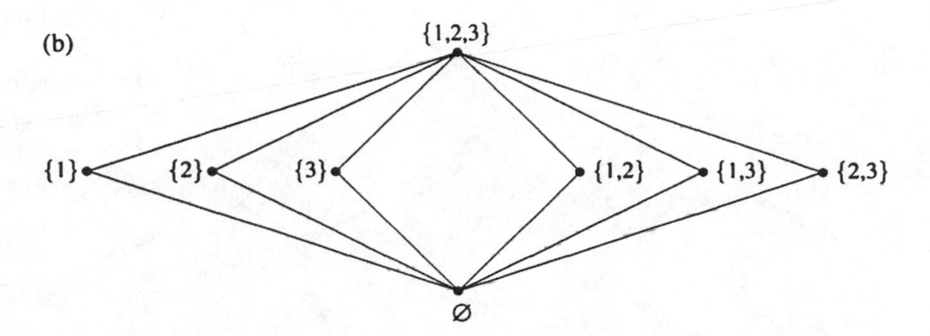

Figure 1. (a) Transition diagram of a quantum-like finite automaton featuring computational complementarity. Input and output symbols are separated by a comma, while arrows indicate transitions. (b) Hasse diagram of its propositional structure. Lower elements imply higher ones if they are connected by edge(s).

(I present you the black box), you may choose which one of the potential and mutually exclusive observables is actually measured (you still have no influence on the particular outcome of the measurement result). This is the precise meaning of the term complementarity mentioned before.

From any partition, we can construct the Boolean propositional calculus that is obtained if we identify its atoms with the elements of the partition. We then "paste" all Boolean propositional calculi (sometimes called subalgebras or blocks) together. This is a standard construction in the theory of orthomodular posets [51–54]. In the above example, we arrive at a form of non-Boolean lattice whose Hasse diagram MO_3 is of the "Chinese-lantern" type shown in Figure 1b.

Let us go still a little bit further and ask which automaton games of the foregoing sort people can actually play. This requires the systematic investigation of all possible non-isomorphic automaton propositional structures, or, equivalently, partition logics [19, 55–56]. Figure 2 shows the Hasse diagrams of all nonisomorphic four-state automaton propositional calculi.

Recall that the method introduced here is not directly related to diagonalization and is a second, independent, source of undecidability. It is already realizable at an elementary pre-diagonalization level,

Figure 2. Variations of the complementarity game up to four automaton states.

i.e., without the requirement of computational universality or its arithmetic equivalent. The corresponding machine model is the class of finite automata.

Since any finite state automaton can be simulated by a universal computer, complementarity is a feature of sufficiently complex deterministic universes, as well. To put it pointedly, if the physical universe is conceived as the product of a universal computation, then complementarity is an inevitable and necessary feature of the perception of intrinsic observers. It cannot be avoided. Conversely, any computation can be realized by a sufficiently complex finite automaton. Universal networks can be formed by serial and parallel composition of finite automata. Therefore, the class of all complementary games is unique and robust, encompassing all possible deterministic universes.

What does all this have to do with intrinsic observers? Well, during the game one is not allowed to "open up" the black box and look inside. Equivalently, one is not allowed to make identical copies of the black box. Both, the "opening up" operation, as well as copying, represent actions that are only accessible to "God-like" external observers, but not to intrinsic ones living in their "Cartesian prison." This is similar to the situation in quantum mechanics. Copying quantum information (unless it is purely classical) or other one-to-many operations is impossible. One cannot, for instance, produce two identical copies of an electron or of a photon [57].

The complementarity game described above shows strong similarities to quantum mechanical systems (in two-dimensional Hilbert space). Indeed, if we could let the black box "shrink" down to a point, we would obtain an analog of an electron or photon, at least for spin or polarization measurements in three different directions. Suppose we want to measure the spin direction of an electron at some angle φ. We can do this by a Stern-Gerlach device oriented in that particular direction. According to the probabilistic interpretation of the wave function, this measurement makes impossible measurements of the spin components in other directions. It may not be legitimate to state that such information [42, 58]) exists independent of the particular act of measurement. Indeed, the propositional structure of three spin measurements along three different angles is identical to the one drawn in Figure 1b.

But there is a difference between the "true" quantum particle and its black box-cousin. Whereas the potential observables associated with the physical particle are supposed to be defined in a *continuum* of directions, the observables associated with the black-box particle can only be generalized to a *countable* number of directions. From a practical point of view, such differences cannot be observed and are therefore operationally irrelevant [20, 59].

Mindboggling Features Beyond Quantum Mechanics

Even to specialists, quantum mechanical effects appear mindboggling [60]. Amazingly enough, the complementarity game based on automata outdoes quantum mechanics in weird behaviors. Take, for example, the complementarity game with the automaton drawn in Figure 3a. Inputting the two-symbol sequence 00 decides between the automaton states 1 and 2 and 3 or 4. The resulting partition is $v(00) = \{\{1\}, \{2\}, \{3, 4\}\}$. Presenting the two-symbol input sequence 10 decides between the automaton states 1 or 2 and 3 and 4. The resulting partition is $v(10) = \{\{1, 2\}, \{3\}, \{4\}\}$. By pasting these two blocks together, we obtain the propositional structure represented in Figure 3b.

This complementarity game has several peculiar features. It is no lattice because the supremum and infimum are not uniquely definable. The "implication" is not transitive either, because $1 \to 1 \vee 2$ requires input 00 and $1 \vee 2 \to 1 \vee 2 \vee 3$ requires input 10, whereas $1 \to 1 \vee 2 \vee 3$ cannot be realized by any experiment.

It would be nice if some day experimenters were to find physical systems that behave in this way. Then, of course, we would have to abandon quantum mechanics and adopt some new theory of the complementarity game.

VII. Is It Possible to Predict Surprises?

The program to study relative limits of knowledge can be attacked from two opposite positions. On the one hand, it may be objected that there are no *real* unknowables, because everything is strictly causal. On the other hand, it may be argued that undecidability in physics is a trivial matter of fact and must be accepted without any further discussion.

The first, rationalistic, position is based on the assumption that the world is totally and in all of its aspects conceivable and predictable by rational (human) thought. Laplace's demon [61] is a metaphor for this position. Indeed, to many physicists, undecidability and unpredictability are everyday phenomena. They encounter a problem that they cannot solve or ask questions they cannot answer. Yet, they would hardly regard this experience as an indication that there is something out there that

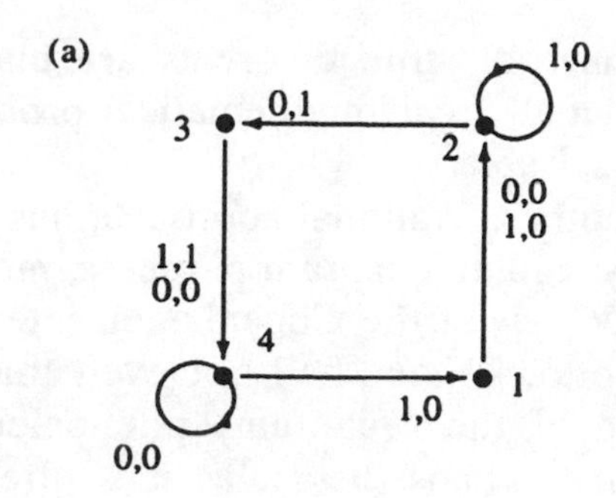

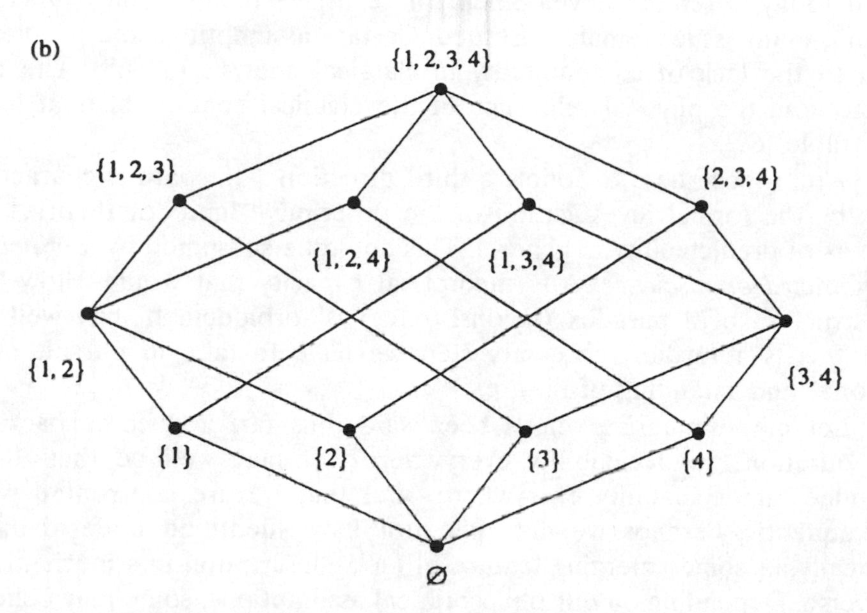

Figure 3. (a) Complementarity game featuring weirder-than quantum properties; (b) Hasse diagram of its propositional structure.

is intrinsically undecidable. It would probably be fair to state that—with the notable exceptions of chaos and quantum theory—most physicists regard undecidable statements not as fundamentally unknowable, but as a challenge and a possibility for future knowledge.

I believe this is a very fruitful and progressive point of view.

Analogously, some mathematicians tend to think of Gödel's incompleteness theorems as artifacts. To them, Gödelian sentences appear curious, even dubious, and explicitly constructed for their purpose. Despite proofs that "almost all" true theorems are undecidable [8], such mathematicians feel that all "real" mathematical problems (i.e., the ones they bother with) *are* solvable.

In this century, another, irrational approach, has been widespread—despite reservations by leading quantum pioneers, most notably Einstein and Schrödinger [25–26]—with the Copenhagen interpretation of quantum mechanics. Its motto, "Thou shall not even think of any rational, realistic understanding of the (quantum) phenomena" contributed to the *mysterium quanticum.* Chaos theory, as it is often called, has given irrationality additional impetus. Already in 1889, Poincaré suggested that certain n-body problems may turn out to be impossible to solve [62]. Even today, after the development of recursive (computable) function theory, many issues remain unsettled. Certain assumptions and problems lead to the lack of computability in classical analysis [63–68]. The necessity and the physical relevance of the classical continuum is at least debatable [59].

We propose here to follow a third direction. This path is characterized by the formal investigation of the descriptive limits of theories, as well as of predictability in general. This could be subsumed by a *principle of restricted omniscience:* Any theoretical capacity that would allow the construction of a paradox (inconsistency) is forbidden. It may well be that this is a further, necessary step we have to take in pursuit of a rational understanding of nature.

Let me summarize what's been said thus far, and come back to the question, "undecidability everywhere?" It may well be that there is indeed undecidability everywhere, and that we are confronted with it frequently. Perhaps we may just not have identified undecidability correctly, as some emerging feature of (self-)description in a mechanistic universe. Depending on our philosophical assumptions, some may believe that the everyday unknowns are either manifestations of some basic randomness, a sort of "chaos" underlying nature, or, on the contrary, that they are simply an artifact of our limited knowledge and power to implement that knowledge. We may also realize that there is still another possibility, having to do with the fact that, informally stated, self-knowledge is necessarily incomplete.

Notes and References

[1] Gödel, K. *Monatshefte für Mathematik und Physik,* 38 (1931), 173. (English translation in Gödel, K. *Collected Works–Volume I: Publications*

1929–1936. S. Feferman et al., eds. Oxford University Press, Oxford, 1986.)

[2] Tarski, A. "Der Wahrheitsbegriff in den Sprachen der deduktiven Disziplinen", in *Akademie der Wissenschaften in Wien, Mathematisch-naturwissenschaftliche Klasse, Anzeiger,* 69 (1932), 24.

[3] Turing, A. M. *Proc. London Math. Soc.* 2, (1936–37), 42, 230 (reprinted in Davis, M. *Computability & Unsolvability.* McGraw-Hill, New York, 1958.)

[4] Chaitin, G. J. *Information, Randomness and Incompleteness,* 2d ed. World Scientific, Singapore, 1990; see also, *Algorithmic Information Theory.* Cambridge University Press, Cambridge, 1987; *Information-Theoretic Incompleteness.* World Scientific, Singapore, 1992.

[5] Calude, C. *Information and Randomness – An Algorithmic Perspective.* Springer, Berlin, 1994.

[6] Casti, J. L. *Beyond Belief: Randomness, Prediction and Explanation in Science.* CRC Press, Boca Raton, Florida, 1990; see also, *Searching for Certainty: What Scientists Can Know about the Future.* William Morrow & Company, New York, 1991.

[7] Rucker, R. *Infinity and the Mind.* Birkhäuser, Boston 1982; reprinted by Bantam Books, New York, 1986.

[8] Calude, C., H. Jürgensen and M. Zimand. "Is Independence an Exception?" *Appl. Math. Comput.,* 66 (1994), 63–76.

[9] Hilbert, D. "Über das Unendliche," *Math. Annalen,* 95 (1926), 161–190.

[10] In a double slit experiment, the wave function Ψ, which is supposed to represent a complete theoretical representation, can indeed be in a coherent superposition $\Psi = \psi(1) + \psi(2)$ of the amplitude for the particle to go through slit 1 and slit 2. Classically, 1 and 2 exclude each other.

[11] Frank, P. *Das Kausalgesetz und seine Grenzen.* Springer, Vienna, 1932.

[12] R. J. Boscovich, *De spacio et tempore, ut a nobis cognoscuntur.* (Vienna, 1755); English translation in *A Theory of Natural Philosophy,* ed. by J. M. Child, Open Court, Chicago, 1922; reprinted by MIT Press, Cambridge, MA, 1966, pp. 203–205.

[13] Toffoli, T. "The Role of the Observer in Uniform Systems", in *Applied General Systems Research,* G. Klir, ed. Plenum Press, New York, 1978.

[14] Svozil, K. "On the Setting of Scales for Space and Time in Arbitrary Quantized Media," Lawrence Berkeley Laboratory, preprint LBL–16097, May 1983.

[15] Svozil, K. *Europhysics Letters,* 2 (1986), 83.

[16] Svozil, K. *Il Nuovo Cimento,* 96B (1986), 127.

[17] Rössler, O. "Endophysics", in *Real Brains, Artificial Minds,* J. L. Casti and A. Karlqvist, eds. North-Holland, New York, 1987, p. 25.

[18] Rössler, O. *Endophysics, Die Welt des inneren Beobachters.* Merwe Verlag, Berlin, 1992.

[19] Svozil, K. *Randomness & Undecidability in Physics.* World Scientific, Singapore, 1993.

[20] Bridgman, P. W. "A Physicist's Second Reaction to Mengenlehre," *Scripta Mathematica,* 2 (1934), 101–117; 224–234 (cf. Landauer, R. "Advertisement For a Paper I Like," in *On Limits,* ed. by J. L. Casti and J. F. Traub, Santa Fe Institute Report WP 94–10–056, Santa Fe, NM, 1994, p. 39. See also Bridgman, P. W. *Reflections of a Physicist.* Philosophical Library, New York, 1950.

[21] Putnam, H. *Reason, Truth and History.* Cambridge University Press, Cambridge, 1981.

[22] Svozil, K. "Extrinsic-Intrinsic Concept and Complementarity", in *Inside Versus Outside*, ed. by H. Atmanspacher and G. J. Dalenoort, eds. Springer, Berlin, 1994, pp. 273–288.

[23] Rasetti, M. *Chaos, Solitons & Fractals,* 5 (1995), 133–138.

[24] Bernstein, J. *Quantum Profiles.* Princeton University Press, Princeton, 1991, pp. 140–141, and J. A. Wheeler, private communication.

[25] *Quantum Theory and Measurement,* J. A. Wheeler and W. H. Zurek, eds., Princeton University Press, Princeton, 1983.

[26] Jammer, M. *The Philosophy of Quantum Mechanics.* Wiley, New York, 1974.

[27] The Bible contains a passage, which refers to Epimenides, a Crete living in the capital city of Knossus: *"One of themselves, a prophet of their own, said, 'Cretans are always liars, evil beasts, lazy gluttons.' "*—St. Paul, Epistle to Titus I (12-13). For more details, see Anderson, A. R. "St. Paul's epistle to Titus," in *The Paradox of the Liar,* ed. by R. L. Martin, Yale University Press, New Haven, 1970.

[28] Rogers, H. *Theory of Recursive Functions and Effective Computability.* McGraw-Hill, New York 1967.

[29] Odifreddi, P. *Classical Recursion Theory.* North-Holland, Amsterdam, 1989.

[30] Popper, K. R. *The British Journal for the Philosophy of Science,* 1, (1950), 117, 173.

[31] Wolfram, S. *Cellular Automata and Complexity, Collected Papers.* Addison-Wesley, Reading, MA, 1994.

[32] Moore, D. *Phys. Rev. Lett.,* 64 (1990), 2354.

[33] da Costa, N. C. A. and F. A. Doria, *Foundations of Physics Letters,* 4 (1991), 363; see also da Costa, N. C. A. and F. A. Doria, *International Journal of Theoretical Physics,* 30 (1991), 1041.

[34] Rado, T. *Bell Sys. Tech. J.,* May 1962, 877.

[35] Chaitin, G. J. *Journal of the Assoc. Comput. Mach.,* 21 (1974), 403.

[36] Chaitin, G. J. "Computing the Busy Beaver Function," in *Open Problems in Communication and Computation,* ed. by T. M. Cover and B. Gopinath, Springer, New York, 1987, p. 108 (reprinted in [4]).

[37] Dewdney, A. K. *Scientific American,* 251, 10 July 1984.

[38] Brady, A. H. "The Busy Beaver Game and the Meaning of Life," in *The Universal Turing Machine. A Half-Century Survey,* ed. by R. Herken, Oxford University Press, Oxford, 1988, p. 259.

[39] Von Neumann, J. *Theory of Self-Reproducing Automata,* A. W. Burks, ed. University of Illinois Press, Urbana, 1966.

[40] Gold, E. M. *Information and Control,* 10 (1967), 447.

[41] Peres, A. *Quantum Theory: Concepts & Methods.* Kluwer Academic Publishers, Dordrecht, 1993.

[42] *Quantum Theory and Measurement,* J. A. Wheeler and W. H. Zurek, eds., Princeton University Press, Princeton, 1983.

[43] Schrödinger, E. *Naturwissenschaften,* 23, (1935), 807; 823; 844 (English translation in *Quantum Theory and Measurement,* J. A. Wheeler and W. H. Zurek, eds., Princeton University Press, Princeton, 1983, p. 152.

[44] Cohen, D. W. *An Introduction to Hilbert Space and Quantum Logic.* Springer Verlag, New York, 1989.

[45] Wright, R. "Generalized Urn Models", *Found. Phys.,* 20 (1990), 881–903.

[46] D. Aerts, "Example of a macroscopic classical situation that violates Bell inequalities." *Lettere al Nuovo Cimento.* 34 (1982), 107–111.

[47] Moore, E. F. "Gedanken Experiments on Sequential Machines," in *Automata Studies,* C. E. Shannon and J. McCarthy, eds., Princeton University Press, Princeton, 1956.

[48] D. Finkelstein, S. R. Finkelstein. "Computational Complementarity." *Inter. J. Theor. Phys.,* 22 (1983), 753–779.

[49] R. Landauer, "Irreversibility and Heat Generation in the Computing Process," *IBM J. Res. Dev.,* 5, (1961), 183–191; reprinted in *Maxwell's Demon,* H. S. Leff and A. F. Rex, eds., Princeton University Press, Princeton, 1990, pp. 188–196; see also, "Fundamental Physical Limitations of the Computational Process; an Informal Commentary," in *Cybernetics Machine Group Newsheet,* 1/1/87; "Computation, Measurement, Communication and Energy Dissipation," in *Selected Topics in Signal Processing,* S. Haykin, ed., Prentice-Hall, Englewood Cliffs, NJ, 1989, p. 18; *Physics Today,* 44, (May 1991), 23; "Zig-Zag Path to Understanding," *Proceedings of the Workshop on Physics and Computation PHYSCOMP '94,* IEEE Press, Los Alamitos, CA, 1994, pp. 54–59.

[50] Bennett, C. "Logical Reversibility of Computation," *IBM J. Res. Dev.* 17 (1973), 525–532. Reprinted in: *Maxwell's Demon*, ed. by H. S. Leff and A. F. Rex, Princeton University Press, 1990, pp. 197–204.

[51] Kalmbach, G. *Orthomodular Lattices.* Academic Press, New York, 1983); see also, *Measures and Hilbert Lattices.* World Scientific, Singapore, 1986; *Foundations of Physics,* 20 (1990), 801.

[52] Pták, P. and S. Pulmannová, *Orthomodular Structures as Quantum Logics.* Kluwer Academic Publishers, Dordrecht, 1991.

[53] Piziak, R. "Orthomodular Lattices and Quadratic Spaces: A Survey," *Rocky Mountain Journal of Mathematics,* 21 (1991), 951.

[54] Navara, M. and V. Rogalewicz, *Math. Nachr.,* 154 (1991), 157.

[55] Schaller, M. and K. Svozil, "Partition logics of automata." *Il Nuovo Cimento,* 109 B (1994), 167–176; see also, "Automaton partition logic versus quantum logic," *Inter. J. Theor. Phys.,* 34 (1995), 1741–1750; "Automaton logic", *Inter. J. Theor. Phys.,* in print.

[56] Svozil, K. and R. R. Zapatrin, "Empirical Logic of Finite Automata: Microstatements versus Macrostatements", submitted.

[57] Herbert, N. *Foundation of Physics* 12 (1982), 1171; see also, Wooters, W. K. and W. H. Zurek, *Nature,* 299 (1982), 802; Milonni, P. W. and M. L. Hardies, *Phys. Lett.,* 92A (1982), 321; Glauber, R. J. "Amplifiers, Attenuators and the Quantum Theory of Measurement," in *Frontiers in*

Quantum Optics, E. R. Pikes and S. Sarkar, eds. Adam Hilger, Bristol 1986; Caves, C. *Phys. Rev.,* D26 (1982), 1817.

[58] Peres, A. *Am. J. Phys.,* 46 (1978), 745.

[59] Svozil, K. "Set Theory and Physics", *Foundations of Physics,* in print.

[60] Greenberger, D. B., M. Horne and A. Zeilinger. *Physics Today,* 46, (August 1993), 22.

[61] P. S. Laplace, *Théorie analytique des probabiletés. Introduction*, in *Oeuvres de Laplace,* 7, (1847), 6.

[62] Poincaré, H. "Sur le problème des trois corps," *Acta Mathematica,* 13 (1890), 5–271.

[63] Specker, E. *Dialectica,* 14 (1960), 175; see also, Kochen, S. and E. P. Specker, "The Calculus of Partial Propositional Functions", in *Proceedings of the 1964 International Congress for Logic, Methodology and Philosophy of Science, Jerusalem.* North Holland, Amsterdam, 1965, p. 45–57; Kochen, S. and E. P. Specker, "Logical Structures Arising in Quantum Theory," in *Symposium on the Theory of Models, Proceedings of the 1963 International Symposium at Berkeley.* North Holland, Amsterdam, 1965, p. 177–189; Kochen, S. and E. P. Specker, *Journal of Mathematics and Mechanics,* 17 (1967), 59 (reprinted in Specker, E. *Selecta.* Birkhäuser Verlag, Basel, 1990); Erna Clavadetscher-Seeberger, *Eine partielle Prädikatenlogik.* Dissertation, ETH, Zürich, 1983; Mermin, D. *Rev. Mod. Phys.* 65 (1993), 803.

[64] Wang, P. S. "The Undecidability of the Existence of Zeros of Real Elementary Functions", *J. Assoc. Comput. Mach.* 21 (1974), 586–589.

[65] Kreisel, G. "A Notion of Mechanistic Theory," *Synthese,* 29 (1974), 11–26.

[66] Ştefănescu, D. *Mathematical Models in Physics.* University of Bucharest Press, 1984 (in Romanian).

[67] Pour-El, M. and I. Richards, *Computability in Analysis and Physics.* Springer Verlag, Berlin, 1989. For a critique, see Bridges, D. S. "Constructive Mathematics and Unbounded Operators – A Reply to Hellman," *J. Phil. Logic,* in press.

[68] Calude, C., D. I. Campbell, K. Svozil and D. Ştefănescu. "Strong Determinism vs. Computability," in *The Foundational Debate: Complexity and Constructivity in Mathematics and Physics,* W. DePauli-Schimanovich, E. Köhler and F. Stadler, eds., Kluwer, Dordrecht, 1995.

Chapter 10

ON REALITY AND MODELS

Joseph F. Traub

I. Introduction

Recently, I heard a researcher present a colloquium on computational aspects of protein-folding. Although this man was obviously an expert on the topic, he casually mentioned in passing that, of course, "protein-folding is NP-complete."

Protein-folding is a biological process that nature performs swiftly. One question that scientists would like to answer is: Given a linear sequence of amino acids, into what three-dimensional configuration will the sequence fold? Experience to date is that this process is very difficult to simulate on the most powerful supercomputers. Fraenkel (1993) proved that a particular mathematical model (minimal energy) of protein-folding is NP-complete in the Turing machine model of computation.

Note that four worlds come into play here; see Figure 1. Above the horizontal line are two real worlds; the world of biological phenomena and the computer world, where simulations are performed. These are worlds of atoms and electrons. Below the horizontal line are two formal models: a mathematical model of the biological phenomenon and a model of computation. In the formal models, representations are in bits.

The mathematical model is an abstraction of the natural world while the model of computation is an abstraction of the computer world. We get to select both of these abstractions and the next section will be devoted to a discussion of these choices.

Discussions of multiple worlds may also be found in Traub and Woźniakowski (1991), Traub (1992), Jackson (1994) and Casti (1996).

The statement "protein-folding is NP-complete" co-mingles a real-world phenomenon with formal models. This is a not uncommon shortcut but if we are to make progress on a theory of scientific limits, it will be

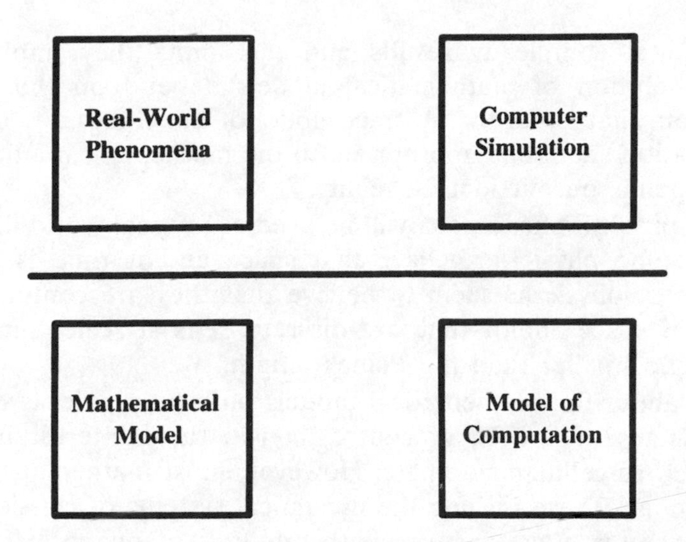

Figure 1. Four worlds.

important to keep the distinction between reality and models clear.

What is the current situation with respect to protein-folding in these four worlds?

- *Protein-folding:* Nature does it fast.
- *Computer simulation:* Protein folding cannot be done on the fastest computers.
- *Formal models:* A particular model (minimal energy) has been proven NP-complete in the Turing machine model of computation. Most experts believe that NP-complete problems are computationally intractable.

After we have built our understanding of some of the issues regarding mathematical and computer models we will explore, in the concluding section, the dissonance among nature, simulation, and models.

I'll summarize the remainder of the chapter. In the next section I will discuss formal models focusing on models of computation. In Section III the intrinsic difficulty of solving a mathematical model, as measured by its computational complexity, will be discussed. In the concluding section I will return to protein-folding and apply what we've learned to present some reasons for the dissonances in our current state of knowledge.

II. Formal Models

Computational complexity results and the limits they imply for the computer solution of mathematical models depend on the model of computation; that is, on the abstract model of the computer. The model of computation should be appropriate to the mathematical model, which in turn depends on our idea of reality.

Here, physical phenomena will be used as my real world illustration. Although some physicists believe that space and/or time is ultimately discrete, most physicists seem to believe that they are continuous. Furthermore, if space and/or time are discrete, it is at scales many orders of magnitude smaller than the Planck length.

What about the mathematical models built by physicists or applied mathematicians? There is, of course, considerable interest in discrete models such as cellular automata. However, most mathematical models are continuous. These include the dynamical systems of classical physics and the operator equations and path integrals of quantum mechanics. That is, in their mathematical models, physicists use number fields such as the real and complex numbers. For simplicity I will refer only to the reals in what follows.

It is well understood that the real numbers are an abstraction. That is, it would take an infinite number of bits to represent a single real number; an infinite number of bits are not available in the universe. Real numbers are utilized because they are a powerful and useful construct.

Let us accept that today continuous models are central to mathematical physics and that they will continue to occupy that role for at least the foreseeable future. But the computer is a finite-state machine. *What should we do when the continuous mathematical model meets the finite-state machine?*

I will compare and contrast two models of computation: the Turing machine and the real-number model. In the interest of full disclosure I want to tell you that I've always used the real-number model in my work but will do my best to present balanced arguments. I will assume the reader is familiar with the Turing machine as the abstraction of a digital computer. In the real-number model we assume that we can store and perform arithmetic operations and comparisons on real numbers exactly and at unit cost. Of course, this is an abstraction and the test is how useful and close the abstraction is to reality.

The real-number model has a long history. It was used for polynomial evaluation (Ostrowski (1954)), in optimal iteration theory (Traub (1964)), algebraic complexity (Borodin and Munro (1975)), information-based complexity (Traub, Wasilkowski and Woźniakowski (1988)), and in continuous combinatorial complexity (Blum, Shub, and Smale (1989)). See also Moore (1995), for recursion theory on the reals and a chaotic dynamical systems approach surveyed by Siegelmann (1995).

What are the pros and cons of these two models? I'll begin with the pros of the Turing machine model.

The attraction of the Turing machine is its simplicity and economy of description. Turing's definition of computability is equivalent to other definitions, and according to the Church-Turing thesis it may be considered a universal definition of computability. See, however, Moore (1995) and, especially Siegelmann (1995), who claims her model is a "super-Turing" machine.

I'll turn to the cons of the Turing machine model. I believe it is not natural to use the discrete model in conjunction with continuous mathematical models. Furthermore, estimated running times are not predictive of scientific computation on digital computers.

I'll move now to the pros of the real-number models. Many mathematical models in physics, and generally in science and engineering, are continuous and use the real number system. For such formulations it seems natural to also use the real numbers in the model of computation.

For example, investigation of the computational complexity of path integrals has recently been initiated; see Wasilkowski and Woźniakowski (1995). The real-number model is used; I believe a Turing machine model would not be natural.

Most scientific computation use finite-precision, floating point arithmetic. Modulo stability, computational complexity in the real-number model is the same as for finite precision floating point. Therefore, the real-number model is predictive of running times for scientific computation.

The final pro that I'll mention here is that by using the real-number model one has at hand the full power of continuous mathematics. Here's just one example of the significance of that. There has been considerable interest in the physics community in the result that there exist differential equations with computable initial conditions and non-computable solutions. (Whether physicists should be concerned about non-computability is an issue that I will take up in another paper.) This follows from a theorem on ill-posed problems established by Pour-El and Richards. They use computability theory to establish their result and devote a large portion of a monograph, Pour-El and Richards (1988), to develop the mathematical underpinnings and to prove the theorem.

An analogous result on ill-posed problems has been established using information-based complexity, which relies on the real-number model. (Information-based complexity will be discussed in Section IV.) The proof takes about one page; see Werschulz (1987). More importantly, in information-based complexity it is natural to consider the average case. It was recently shown that every ill-posed problem is well-posed on the average for every Gaussian measure; see Traub and Werschulz (1994) for a survey. There is no corresponding result using computability theory. The theme of average behavior will play a prominent role in the final two sections.

An eloquent argument for the real-number model is given in the "Manifesto" by Blum, Cucker, Shub, and Smale (1995). They write "Our point of view is that the Turing model ... is fundamentally inadequate for giving a foundation to the theory of modern scientific computation."

The con of using the real-number model is that it would be attractive to use a finite-state model for a finite-state machine.

The pros and cons of the Turing machine and real-number models are summarized in Table 1.

Table 1. Pros and Cons of Two Models.

– Turing Machine Model –

Pro:

- Simple, robust

Con:

- Not predictive for scientific computation

– Real-Number Model –

Pro:

- "Natural" for continuous mathematical models
- Predictive for scientific computation
- Utilizes the power of continuous mathematics

Con:

- Attractive to use finite-state model for finite-state machine

Some of my colleagues are uncomfortable with the use of the real-number model because they believe that both the mathematical models and the model of computation should be finite. See, for example, the brief notes by Casti, Jackson, and Landauer in Casti and Traub (1994).

Note that the Turing machine model is *not* finite since it uses an unbounded tape. I would characterize the Turing machine as discrete but unbounded. Then, why not use a finite model of computation? There are such models (for example, circuit models and linear bounded automata), but they are special purpose. See Table 2 for the distinctions.

Table 2. Finite and Unbounded Models.

Finite Models:

- Circuits
- Linear bounded automata

Unbounded Models:

Discrete

- Turing Machine

Continuous

- Real-number Model

The idea of using only finite mathematical and computational models is certainly attractive. We'll have to wait and see if scientists succeed in building them.

III. Computational Complexity

Computational complexity measures the minimal computational resources required to solve a mathematically-posed problem. For brevity, I'll often use "complexity" for the remainder of this paper. I'll comment on this informal definition:

- Consider all possible algorithms for solving a problem; those that are known and those existing only in principle. The complexity is the minimal cost over all possible algorithms.
- In Traub (1991), I suggest how ideas analogous to those used in complexity might be used to prove limits to scientific knowledge. I will not pursue that theme here.
- The complexity may be regarded as measuring the *intrinsic* difficulty of a mathematically posed problem.

- The "computational resources" may be time, memory, area on a chip, etc. In this paper the resource will always be time. Then the complexity is the minimal time required to solve a problem exactly or to prescribed accuracy.
- The complexity depends on the problem, not on the algorithm for solving it. It also depends on the model of computation and on the guarantee we offer regarding the solution (the "setting"). I'll return to this below.
- Computational complexity is both the difficulty of a problem and the name of a field of study. The meaning is usually clear from context.
- Complexity may be thought of as the "thermodynamics of computation" with intrinsic limits on what any heat engine can do replaced by limits on what any algorithm can do. See Packel and Traub (1987).

Computational complexity comes in various flavors. The structure is shown schematically in Figure 2. The top node in this tree is all of complexity. This may be divided into discrete and continuous complexity.

The node labeled *discrete* represents discrete combinatorial problems. Typical here is the well-known Traveling Salesman Problem (TSP). The input is the location of *n* cities; these locations are usually represented with a finite number of bits. The input specifies a single TSP; the information is *complete.*

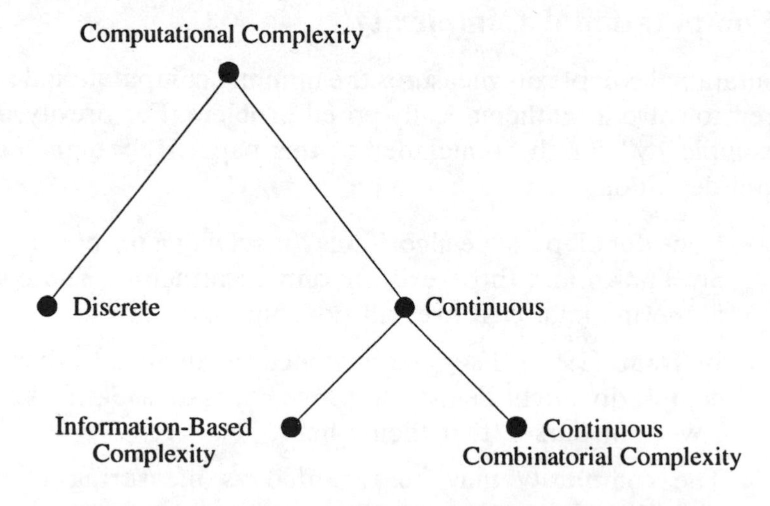

Figure 2. Schema of computational complexity.

Continuous complexity may be divided into two parts; information-based complexity (IBC), and continuous combinatorial complexity. Typical problems of IBC are multivariate and path integration. Most integration problems that occur in practice have to be solved numerically. The mathematical input is the integrand but the information available for solving the problem consists of a finite number of integrand evaluations. This information usually does not specify an integrand uniquely; the information is *partial.*

Finally, a typical problem of continuous combinatorial complexity is 4-satisfiability; does a system of quartic polynomial equations have a real zero? The input to this problem consists of the coefficients, taken as real numbers.

Table 3 distinguishes among these three areas of computational complexity with respect to the model of computation and available information. Note that combinatorial complexity, whether discrete or continuous, makes the same assumption about information. The difference is that discrete combinatorial complexity uses the Turing machine or an equivalent model, whereas continuous combinatorial complexity uses the real-number model. Note that IBC and continuous combinatorial complexity use the real-number model but make opposite assumptions about information.

I will briefly indicate results, starting with combinatorial complexity. How does the complexity grow with the size of the input? For example, in TSP, the size of the input is the number of cities, n. Typically, we do not know the complexity of combinatorial problems. We don't even know if the complexity grows polynomially or exponentially with the size of the input.

If the complexity grows polynomially we say the problem is *tractable;* if the growth is superpolynomial, e.g., exponential, we say it is *intractable.*

Since we don't know the complexity of combinatorial problems we have to settle for a complexity hierarchy. Perhaps the hierarchy collapses, at least partially. The famous conjecture $P \neq NP$ states that at least a portion of the hierarchy does not collapse; see, for example, Papadimitriou (1994).

Today we do not know if TSP is tractable or intractable. Most experts believe $P \neq NP$ and that TSP is therefore intractable. But that remains only a conjecture.

What we do know is that many combinatorial problems are equivalent from the complexity viewpoint. They are all tractable or all intractable. The "hardest" problem in the class of NP problems, in the sense

Table 3. Flavors of complexity.

	Complexity type		
	Discrete combinatorial	**Continuous combinatorial**	**Information-based**
Model	Turing machine	Real-number	Real-number
Information	complete exact free	complete exact free	partial contaminated priced

of reduction, is said to be NP-complete. Blum, Shub, and Smale (1989) gave a certain formalization of the real-number model, often called a BSS machine. They established that 4-satisfiability is NP-complete over the reals.

We conclude this section with results from information-based complexity. Here we do often have tight bounds on complexity and do not have to content ourselves with a complexity hierarchy. I'll use the diagram of Figure 3 to explain why we can obtain complexity bounds in IBC.

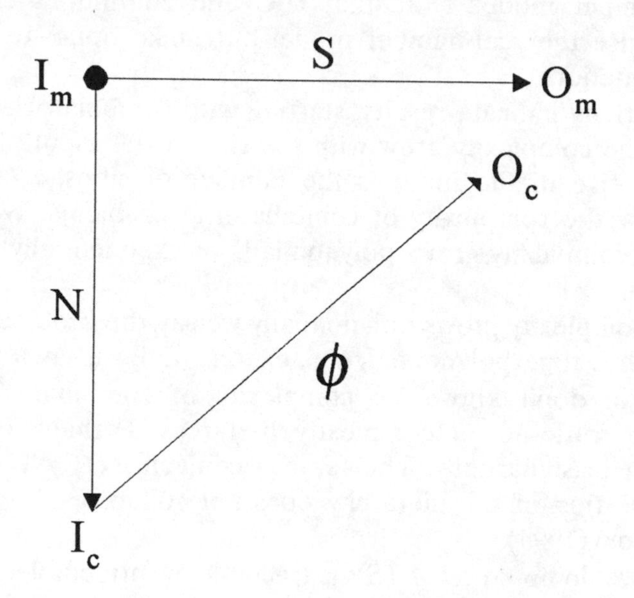

Figure 3. Schema for information-based complexity.

The mathematical problem to be solved is specified by the operator S that maps the mathematical input, I_m, into the mathematical output O_m. This is very general since one can think of all computation as taking inputs into outputs.

Suppose now that the mathematical input is a real multivariate function. Such a function cannot be input to a digital computer. Thus the function has to be replaced by a finite set of numbers, say, evaluating the function at a finite number of points. The operator N maps the mathematical input, I_m, into the computer input I_c. It's crucial that N is a many-to-one operator, i.e., knowing I_c does not give us I_m. Indeed, in IBC there are typically an infinite number of indistinguishable mathematical inputs corresponding to a computer input.

A computer algorithm maps the computer input, I_c, into the computer output O_c. Note that $O_c \neq O_m$. Since N is many-to-one, we can't know which mathematical problem we're solving and therefore can, at best, solve the problem only approximately. Mathematically stated, N composed with ϕ does not commute with S.

Now I can explain why we can often get tight lower and upper bounds on the computational complexity of IBC problems. We can use arguments based on how powerful the information operator has to be. (Indeed, this is why the field is called information-based complexity). See the monograph by Traub, Wasilkowski, and Woźniakowski (1988) for rigorous mathematical formulation and analysis, and Traub and Woźniakowski (1994) for a more informal treatment.

For combinatorial problems the computer input is usually the same as the mathematical input and there are no information-based arguments.

Although IBC is an abstract theory developed over abstract spaces, the typical applications are to multivariate functions. Here is a typical result from IBC. The problem is integration of a function defined on the unit cube in d dimensions. Assume we are in the worst case deterministic setting. That is, we guarantee an error at most ϵ for every integrand in some class of integrands and randomization is not permitted.

Let the class of integrands be continuous; smoothness is not assumed. Then it is easy to see that the complexity is infinite for all ϵ; that is, the problem is *unsolvable.*

Assume next that the class of integrands is once continuously differentiable with uniformly bounded derivatives. Then the complexity is proportional to $(1/\epsilon)^d$. That is, the complexity increases exponentially with dimension and the problem is *intractable.*

For many classes of functions that have fixed smoothness (in the sense of Sobolev), integration is intractable. For the precise result see, for example, Traub and Woźniakowski (1991, 1994).

Thus the integration problem is unsolvable or intractable in the worst case deterministic setting. But this is not an anomaly; typically multivariate continuous problems are intractable.

Since this is a complexity result we can't beat the intractability result by inventing a clever new algorithm. *The only way to possibly break intractability is to weaken the assurance.*

I'll mention two settings with weaker guarantees. One is the Monte Carlo (randomized) setting. The guarantee here is that the expected error, with respect to the distribution on the sample points, is less than ϵ. Then the complexity of integration is proportional to $1/\epsilon^2$, independent of d, even when the class of integrands is only continuous. (Recall that in the worst case deterministic setting this problem was unsolvable).

The second setting is the average case deterministic setting. Assume a Wiener measure on the continuous functions. The guarantee is now that the expected error with respect to the Wiener measure is less than ϵ.

Since this is a deterministic setting, the evaluation points must be given. This was a long-open problem of optimal design solved by Woźniakowski (1991). He established a connection to low discrepancy sequences in number theory and showed that the complexity is proportional to $1/\epsilon$, modulo a polylog factor in $1/\epsilon$. Numerical tests on a problem of mathematical finance involving integration in 360 dimensions (Paskov and Traub (1995)) suggest that evaluation at low discrepancy points may be superior to Monte Carlo for certain problems in mathematical finance.

That completes our brief tour of concepts and results from computational complexity. In the concluding section I'll return to the protein-folding problem, applying what we've learned.

IV. Application to Protein Folding

Now that we are equipped with an arsenal of ideas from computational complexity, I'll return to the issue raised at the beginning of this paper regarding protein-folding. Here's what is known about the current status of this biological problem:

- Nature does it quickly
- We cannot simulate the process on even the most powerful supercomputers

• A particular mathematical formulation is believed to be computationally intractable in a particular model of computation

It seems to me that a natural question is how does the time that nature uses to do protein-folding depend on the length of the sequence of amino acids. Since nature folds proteins very fast and since the length of N, the amino acid string is large, the time is not exponential. Is the time superlinear, sublinear, or even constant, independent of n? It is my understanding from a conversation with Jonathan King that this is a question that experimentalists have not asked.

Note that there are two separate issues here:

• to explain how protein-folding occurs in nature, and

• to ask if we can perform a computer simulation of this process.

A similar dichotomy occurs in vision research. We want to

• understand the human visual system, and

• give machines similar abilities.

Both issues are of interest. Humans have excellent visual and pattern recognition skills due to millions of years of evolution. It has proven very difficult to give computers such abilities. Progress on one issue might help with the other but not necessarily.

Should we be concerned that nature does protein-folding easily while, with our current knowledge, simulation seems hard in practice and theory? Not necessarily, but in the list presented below I will suggest some possible reasons for the difference. First, I'll remind the reader of the current theoretical status. Fraenkel (1993) proved that a minimal energy model based on a discrete graph representation is NP-complete in the Turing machine model. That is, if the sequence of amino acids is of length n and if the conjecture $P \neq NP$ is true, then the problem cannot be solved in running time that is a polynomial in n.

a) Nature has selected for proteins that fold easily.

b) NP-completeness is a worst case theory. Perhaps the solution of the mathematical model is easy on the average but we don't know the prior. Note that nature using selection is one example of an unknown prior.

The average behavior can be totally different from the worst case behavior; in Section III, I used the example of high dimensional integration to show that a problem that is unsolvable or intractable in the worst case can be tractable in the average case. Here is a second example. Before the Karmarkar algorithm, the simplex algorithm was the algorithm of choice for solving linear programming. Although, due to a result of Klee and Minty, the cost of the algorithm was known to be exponential in the worst case, practitioners reported that the cost was a low degree polynomial in the size of the problem. Then Borgwardt (1982) and Smale (1983) independently proved that the average cost of the simplex algorithm is a low degree polynomial; this gave a theoretical explanation of what practitioners had experienced. Thus there is an exponential difference between the worst and average case running time. (Note that I'm careful to talk about *cost* and not complexity, since these are only properties of a particular algorithm.)

c) A minimal energy model is commonly used. Perhaps there are other mathematical models that are not computationally intractable.

d) Fraenkel assumes that the mathematical model is exactly solved but perhaps it's enough to solve it approximately. IBC problems can *only* be approximately solved because the information is partial. Combinatorial problems can often be exactly solved. It is possible for a combinatorial problem that is intractable if an exact answer is demanded to become tractable if an approximate answer suffices. Thus we may *choose* to solve a combinatorial problem approximately. See, for example, Garey and Johnson (1979).

e) The set of inputs must be specified. For some problems, such as TSP, any set of n points in the Euclidean plane can be an input. For other problems, the choice of input set can totally change the problem complexity. I'll illustrate the point with the simple example of univariate integration in the worst case setting. If the class of inputs consists of continuous functions then the problem is unsolvable. If the class of inputs consists of continuous differential functions with uniformly bounded derivative, then the complexity of computing an approximation with error at most ϵ is proportional to $1/\epsilon$.

f) Nature may be using massive parallelism, say, of order 10^{23}. Such parallelism might eventually be provided by quantum computation (see Di Vincenzo (1995) for a recent survey and the references given there), or biological computation (see Adelman 1995), but this is currently highly speculative.

g) The NP-completeness result uses the Turing machine model of computation. Perhaps a different model of computation might be more appropriate. Possibilities are the real-number model or the "super-Turing" model mentioned in Section II.

h) We should not forget that intractability of NP-complete problems is only conjectured.

i) Nature may have ways of cutting the complexity of protein folding. See the last section of Fraenkel (1993) for some examples. Fraenkel gives an excellent general discussion of the ramifications of his NP-completeness result.

How might this affect the dissonance between reality and simulation? I'll consider three possibilities from the above list. Algorithms that guarantee good *average* behavior can be very different from those that guarantee good *worst-case* behavior (Point b). Algorithms that are guaranteed to solve a problem approximately can be very different from those that solve a problem exactly (Point d). Thus weakening the guarantee regarding the solution might lead to algorithms that are much cheaper than those in current use.

Finally, the complexity depends on the set of inputs (Point e). For the protein-folding problem the inputs are linear sequences of amino acids of length n. Any prior knowledge restricting the class of inputs might reduce the complexity.

How hard will simulation of protein-folding be in one to two decades? Opinions among biologists vary. When I was asking scientists in the early nineties for their candidates for very hard problems a number of them mentioned simulation of protein-folding as a candidate. On the other hand, Leroy Hood thought it would be routinely solved in one to two decades.

We will see.

Acknowledgements

I'm indebted to John Casti for numerous conversations on the issues discussed in this paper. In particular, we discussed the "four worlds" of Figure 1. Jonathan King told me about the paucity of knowledge regarding how long it takes proteins to fold, as related in Section IV. I want to thank Lee Segel for a number of valuable conversations on protein-folding. I am grateful to Kathi Selig, Arthur Werschulz, and Henryk Woźniakowski for their comments on the manuscript.

References

[Adelman 94] Adelman, L. M. "Molecular Computation of Solutions to Combinatorial Problems," *Science,* 266 (1994), 1021–1024.

[Blum, Cucker, Shub, Smale 95] Blum, L., Cucker, F., Shub, M. and Smale, S. "Complexity and Real Computation: A Manifesto." Technical Report TR–95–042, International Computer Science Institute, Berkeley, CA, 1995.

[Blum, Shub, Smale 89] Blum, L., Shub, M., and Smale, S., "On a Theory of Computation and Complexity over the Real Numbers: NP-Completeness, Recursive Functions and Universal Machines", *Bulletin of the American MathematicalSociety,*21 (1989), 1–46.

[Borgwardt 82] Borgwardt, K. H. "The Average Number of Steps Required by the Simplex Method is Polynomial", *Zeitschrift fur Operations Research,* 26 (1982), 157–177.

[Borodin and Munro 75] Borodin, A. and Munro, I. *The Computational Complexity of Algebraic and Numeric Problems.* Elsevier, New York, 1975.

[Casti 96] Casti, J. "On the Limits to Scientific Knowledge," *Scientific American,* to appear October 1996.

[Casti and Traub 94] Casti, J. and Traub, J.F. "On Limits", Santa Fe Institute Working Paper, WP–94–10–056, 1994.

[DiVincenzo 95] DiVincenzo, D. P. "Quantum Computation", *Science,* 270 (1995)", 255–261.

[Fraenkel 93] Fraenkel, A. S. "Complexity of Protein-Folding", *Bulletin of Mathematical Biology,*55 (1993), 1199–1210.

[Garey and Johnson 79] Garey, M. R. and Johnson, D. S. *Computers and Intractability.* W. H. Freeman, San Francisco, 1979.

[Jackson 94] Jackson, E. A. "The Second Metamorphosis of Science: A Working Paper", Center for Complex Systems Research, Beckman Institute, U. of Illinois, UIUC–BI–CCSR–94–1, 1994.

[Moore 95] Moore, C., "Recursion Theory on the Reals and Continuous-time Computation", Santa Fe Institute Working Paper, WP–95–09–079, 1995.

[Ostrowski 54] Ostrowski, A. M. "On Two Problems in Abstract Algebra Connected with Horner's Rule", in *Studies Presented to R. von Mises,* Academic Press, New York, 1954, pp. 40–48.

[Packel and Traub 87] Packel E., and Traub, J. F. "Information-based Complexity", *Nature,* 328 (1987), 29–33.

[Papadimitriou 94] Papadimitriou, C. H., *Computational Complexity.* Addison-Wesley, Reading, MA, 1994.

[Paskov and Traub 95] Paskov, S. and Traub, J. F., "Faster Valuation of Financial Derivatives", *Journal of Portfolio Management,* 22 (1995), 113–120.

[Pour-El and Richards 88] Pour-El, M. B. and Richards, J. I., *Computability in Analysis and Physics.* Springer-Verlag, Berlin, 1988.

[Siegelmann 95] Siegelmann, H. T. "Computation Beyond the Turing Limit", *Science,* 268 (1995), 545–548.

[Smale 83] Smale, S., "The problem of the average speed of the simplex method", in *Proceedings of the 11th International Symposium on Mathematics,* Springer-Verlag, 1983.

[Traub 64] Traub, J. F., *Iterative Methods for the Solution of Equations.* Prentice-Hall, Englewood Cliffs, N. J., 1964 (reissued by Chelsea Press, New York, 1982).

[Traub 91] Traub, J. F., "What is Scientifically Knowable?", in *Twenty-Fifth Anniversary Symposium,* School of Computer Science, Carnegie-Mellon University, Addison-Wesley, Reading, MA, 1991, pp. 489–503.

[Traub 92] Traub, J. F., "Can We Prove There are Limits to What is Knowable About the Universe?", Computer Science Department, Columbia University, 1992 (Presented at The Mainichi Shimbun Symposium, December, 1992, Tokyo and Osaka, Japan).

[Traub and Werschulz 94] Traub, J. F. and Werschulz, A. G., "Linear Ill-Posed Problems are Solvable on the Average for All Gaussian Measures", *The Mathematical Intelligencer,* 16, No. 2 (1994), 42–48.

[Traub and Woźniakowski 94] Traub, J. F. and Woźniakowski, H., "Breaking Intractability", *Scientific American,* January 1994, 102–107.

[Traub and Woźniakowski 91a] Traub, J. F. and Woźniakowski, H., "Theory and Applications of Information-Based Complexity". Presented in the 1990 Lectures in Complex Systems at the Santa Fe Institute, Addison-Wesley, Reading, MA, 1991, pp. 163–193.

[Traub and Woźniakowski 91b] Traub, J. F. and Woźniakowski, H., "Information-based Complexity: New Questions for Mathematicians", *The Mathematical Intelligencer,* 13 (1991), 34–43.

[Traub, Wasilkowski, and Woźniakowski 88] Traub, J. F., Wasilkowski, G. W., and Woźniakowski, H., *Information-Based Complexity.* Academic Press, San Diego, CA, 1988.

[Wasilkowski and Woźniakowski 96] Wasilkowski, G. and Woźniakowski, H., "On Tractability of Path Integration", *J. Math. Physics,* to appear 1996.

[Werschulz 87] Werschulz, A. G., "What is the Complexity of Ill-Posed Problems?", *Num. Func. Anal. Optim.*, 9 (1987), 945–967.

[Woźniakowski 91] Woźniakowski, H., "Average Case Complexity of Multivariate Integration", *Bulletin of the American MathematicalSociety,* 24 (1991), 185–194.

Index